To Live On Anyway

By
Ann Hageus

www.AuthorMikeInk.com

ISBN: 978-0-9884468-5-4
Library of Congress Control Number: 2013934315

Originally published in Sweden by Per Holm at Holms Förlag
http://holmsforlag.se - Publish Date, December 2009

Book translated from Swedish to English by Alexandra Kent.

First Published by *AuthorMike Ink*, 9/24/2013

www.AuthorMikeInk.com

AuthorMike Ink and its logo are trademarked by *AuthorMike Ink Publishing*.

Printed in the United States of America

I would like to dedicate this book to Gösta, Julius, Karin, and their daddy, Bertil. And to all the children who lost their lives in the tsunami and their families.

Not just them, but to all kids who are deceased and to those whom they left behind. This book is dedicated to any child who died too young and their family.

PROLOGUE

It should be a perfectly ordinary mother sitting here in a field with her children playing around her one warm day in early June 2006. But it isn't.

It's me who is sitting here and I'm no ordinary mother because one of my children has died. He disappeared in the tsunami on December 26, 2004.

Gösta is dead. Gösta is my son. He was nine years old when he died. He has a brother who is two years older called Julius and a little sister called Karin, who is six years younger. Gösta was the putty that bound Julius and Karin together; he was the middle sibling. He and Julius always had each other and they were very close, but he also enjoyed playing with Karin. He used to play so sweetly, always trying to meet her on her own level. He used to let her win when they wrestled and she was allowed to take control and beat him in their games.

It is late afternoon and the air is beginning to cool. Although I'm sitting on a blanket, I can feel the damp creeping up from the ground and I can smell the faint scent of grass. I pull up my legs and cross one over the other. My left foot lands on my right knee and I can feel how the muscle pulls in my buttock. I'm nowhere near as supple as I used to be. I strain to bend forwards and look at the scar on my right toe. The scar still glows bright red even though one and a half years have passed. When I touch it, it feels slightly numb and sometimes it tingles. They say that time heals wounds – but how?

There are children running around not far from me; I think they're playing tag. After a while I can make out Julius and Karin walking towards me a little way from the other children. They are walking together, as though they are talking to one another.

I feel so warm and content when I see them walking side by side, my children, my wonderful children. It is thanks to them that I can go on living. They are my strength. I feel glad because I want them to be there for one another and I want there to be a bond between them. I can see it now. But it is so excruciatingly painful to see this bond because in it is a gaping abyss. They look so different, a big boy and a little girl. Julius is almost twice as tall as Karin and they seem so ill-matched.

Gösta should have been there between them. The bridge. The link between the tall and the short one. One, two, three.

Now Gösta is dead. But no matter how many times I write it, it seems as though I'll never completely accept that it's true, even though I knew that Gösta was gone only minutes after the wave swept over us. I broke off in the middle of saying his name while I was looking for him. My whole body crumpled and I realised it was over and that there was no point in continuing to call after him. Was that the moment at which he died?

Today's end of school picnic for the six-year-olds in Karin's class is almost over and most people have already gone home.

I remain sitting on my blanket and I watch as Karin and Julius begin to play with the children who are left. Bertil is sitting on another blanket a little way away from me, where he is talking to one of the other parents. I look away and allow my thoughts to roam freely.

Why is it just as terrible to lose a one-year-old, a nine-year-old or a twenty-year-old child?

I think it has to do with the hopes you have when you're expecting a baby. As your belly swells, so does the longing for new life. You have nine months to conjure up dreams and visions about this little creature that is slowing developing inside you. You imagine this tiny dot growing into a complete person, but there is this constant worry that something may not be right or something may happen. The child feels so close inside you but it is also so far away. No one has seen it. No one knows anything about it. It is an unwritten page.

Maybe the worry dissolves when the child is born? It's easier to protect the child then, when you can hold it in your arms and see how it is faring.

But even when the child is still inside you, you begin to imagine him, imagine him getting his first tooth, taking his first step. It's so far away, but you nevertheless know what you have ahead of you. You take it for granted that you've begun a journey that will continue throughout your life. On this journey, your child will begin school, finish school, fall in love, badger you for things, cut ties with his parents and leave home, get a job, maybe start his own family. He may even give you grandchildren.

You begin to prepare yourself to be there, offering love and security to this child, doing your best to make him happy and able to stand on his own two feet and enjoy life.

When you have a child in your belly, it's as though a paved road stretches ahead of you and all you have to do is follow it.

My road lasted only nine years, ten months and six days. And then it ended. I had had time to experience Gösta's first tooth, his first steps, his time at preschool and his entry into primary school.

The rest remains only in my imagination and dreams about Gösta. I can still dream of his graduation from school, his first girlfriend. I can worry about him leaving home to travel the world. I can see him troubled by the injustices of this world and I see others waiting expectantly for what he has to offer. His entire life exists in my dreams. I was permitted to experience some of it in real life, but the rest can only be dreamed.

It makes no difference that he was nine instead of one or twenty when he died. It's all the same and equally horrific. The only difference it makes is in relation to how much I was able to experience in real life and how much can only be dreamed.

I remember Gösta at the St. Lucia celebrations. He was two years old and it was late November. The children were going to practice the annual performances for St. Lucia day. They were all allowed to choose which of the traditional outfits they would wear and all of them decided on one of the outfits appropriate to whether they were a girl or a boy. But not Gösta. He wanted to dress up as Lucia herself and there was no problem because his teachers were so open-minded.

I remember how proud I felt when Gösta sailed in with his long white Lucia dress on and a crown of five burning candles on his head.

He continued to play Lucia each year for the five years he was at preschool.

Gösta was a free spirit right from the start.

Gösta is dead. But what does that mean? I realize that it means I will never be able to see him again. And I'm beginning to understand that it means he is gone even in other ways. I try to behold what has happened as though I'm on a higher plane. How does the emptiness after Gösta look? What is it that is so painful? I have to look at the words and repeat them aloud to myself so as to force myself to comprehend that which is so incomprehensible. I repeat them yet again: Gösta is dead. But this just feels like self-mortification.

Why should I force myself to repeat these cruel words when I have to continue living? It's such a contradiction. If Gösta is dead, of course I can't continue to live, but what choice do I have? How can I relate to this? Maybe it's simply a question of continuing to live. But then why do I have to keep reminding myself that Gösta is dead? Is this a way of blunting the edges of the words until they finally lose all definition and meaning? No, it isn't that. It's a way of making them sink in.

Imagine if I was a little more like an animal that didn't care so much about its offspring, if I was like the lioness I saw on television once who abandoned one of her cubs when it got a thorn in its paw. A thorn in a paw was sufficient reason for the mother to leave the cub to its fate.

The lioness was beating a track over the savannah with her three cubs behind her. One of the cubs had begun to limp. Instead of leaping and bounding along just behind his mother he now began to lag further and further behind. After a while, he was far behind his family. He sat himself down on the ground and began licking his paw but his mother and siblings didn't so much as turn to see what had happened. The little cub got up and tried to run to catch up but he only managed a few steps before the thorn forced him to give up.

The cub never caught up. Instead, the distance between him and his family became greater and greater. Maybe his mother glanced back a couple of times before she disappeared over the crest of a hill with her other two cubs.

Nature is cruel. You have to be perfect to survive. Was there something imperfect about Gösta? No, it's not like that. Nature isn't cruel. Nature simply is. Nature does not pity. Nor does it demand that things should be right or wrong because it sees no consequences and takes no consequences. It doesn't evaluate. It simply does. It is only our interpretation of nature that is cruel because its consequences are cruel for us.

It is we humans who pity, evaluate, and take consequences. The lioness didn't leave her cub because she was evil or because she didn't care about it. She did so because she had no choice.

And I have no choice either. I have to go on; I have to continue living.

Like the lioness that was unable to save her cub, I was unable to save Gösta. And like the lioness, who had to continue with her remaining two young, so must I continue with mine.

PART I
2005

The Flight Home from Singapore

Karin is sprawled out asleep in the seat beside me. She has kicked off the blanket and is lying with her mouth a fraction open. I pull myself up, stretch out my legs under the seat in front of me and bring my hands up behind my neck to massage it with my fingertips. My neck is aching after the nine-hour flight. Karin's little hand has fallen onto my knee and I grasp hold of it. It is so relaxed and small and soft.

It is beginning to get light outside the window and when I lean over Karin to look out I can see that the ground is white with snow and ice. The screen before me says we should be landing at Arlanda airport in half an hour.

We have never been so widely dispersed: Karin and I are way up in an airplane on our way home to Stockholm, Bertil is still in Singapore, Julius is already back in Täby and Gösta is so far away that we will never be able to reach him.

A moment later Karin wakes up, stretches and looks sleepily up at me. And then she remembers, she turns and looks out. "Are we there soon?" she asks eagerly.

"Yes," I say. If we look down maybe we can see where we are.

I lean over Karin once more to look out of the window. Interspersed with the white I can now see green patches that are growing larger as we come closer. And suddenly I can see the approach to Stockholm, the grand mansions paraded along the coastline, a large Finland ferry gliding slowly along beneath us. The city becomes clearer and I point out the boat to Karin and then the Globe Arena. Then, as I see Slussen and feel the plane veering off over Skeppsbron towards Kungsträdgården, my mind becomes crowded with thoughts. Time stands still for a moment.

Stockholm is so beautiful. The sun has just risen behind us and is casting golden, orange rays over the old city. The whole city is glistening with white snow. The sea is frozen but the boats have broken through the ice, leaving grey floes floating in the black furrow behind them. The church towers and roofs are chiselled out against the bright blue of the sky. The calendar says that in a few days, March, the first months of spring will be here. It looks so cold.

Stockholm is bathed in strong bright colour.

The contrast between what I see before me and what I carry inside becomes boundless. It is so magnificent to come back to Sweden after our time in Singapore and experience precisely this view. I never fail to be impressed by the majesty of this city.

But why could we not experience this homecoming together, the whole family?

I remember so clearly how we sat on the plane to Singapore almost two years ago. At the time I thought that this was only going to be a short episode in our lives and I kept reminding myself of how important it would be to enjoy the experience and live in the present while we were there. It would all pass so quickly and then we would be back to our everyday life in Sweden and everything would be as before.

Maybe it even crossed my mind then how we would one day all be sitting on the plane again on our way home thinking how fast time had gone since we moved Singapore. Maybe I sat and smiled at all my memories.

Instead, I am bringing home an unthinkably horrific tragedy.

Gösta will never come back home and nothing will ever again be as it was.

Homecoming

We had bought a house off the internet. Just a few days after the catastrophe we decided that we had to leave Singapore and find a house in Täby, a suburb of Stockholm. I soon found one that required thorough renovation, so in the weeks following our return to Sweden I was fully occupied with overseeing the work. We were staying with Bertil's sister Eva, who lived nearby. Each day, I walked over to the new house to check and discuss what was going on there.

The most important thing was to avoid thinking, to try to establish a foundation to stand on. Julius had only just entered the fifth grade at one school when it was already time for him to shift to the sixth grade at another school. Karin would soon start day care and I was advised to start applying for jobs so that I could become re-registered in the Swedish social insurance system. I was not entitled to any kind of leave. A friend helped me find work with the County Council, but I had to start up my own company in order for them to make payments to me. At the same time, I enrolled with the Employment Office and began attending the course they offered for the unemployed.

Almost every day I spoke to Bertil, who was still in Singapore, to get updates about the process of identifying Gösta. After he returned to Sweden a month later, we continued to discuss the search for Gösta with the Swedish police in Thailand. Before we had left Thailand we had established good contact with a policeman called Jonas.

Täby April

On 28 April, a Tuesday morning, as I am walking down a long corridor in Huddinge Hospital, my mobile phone rings. It is Bertil ringing to say what I have been dreading yet waiting for for four months and two days. Gösta has been found – dead.

I put down the phone and begin to feel something that nowadays is all too familiar: a feeling of being enveloped by numbness. I stop breathing and begin to walk as cautiously as though I was in a shop full of priceless porcelain that might break if I make a false move.

When I arrive at the home of Bertil's sister, where we are still living, the policeman Jonas is sitting by the kitchen table. He stands up and comes forward to hug me. I allow him to hold me without words and then I go to the sink and make myself a cup of tea. It is like meeting an old friend seeing him here. But the contradiction between giving him a welcoming 'hello' and knowing why he has come means that I am unable to say anything to him. "I'm so sorry," he says quietly.

I can only shake my head. I go slowly around the table and sit down. Bertil is already sitting at the table and I can see traces of tears on his cheeks. The air is laden with unspoken words.

Jonas sits back down on his chair and breaks the silence. "We had a clear track to follow for Gösta relatively early on because of his yellow speedo swim trunks. It turned out that he was the only person wearing that kind of swimwear. But this was not enough to confirm his identity. We had to have fingerprints, x-rays and a dental status. It was a great help that you were able to send his dental records and the x-rays of his broken leg.

But it was your help with the fingerprints that was crucial."

"It was thanks to the fact that Bertil stayed in Singapore that you got those," I can hear my voice saying.

"I think that was the hardest time of my entire life, being left behind alone in Singapore after you'd all gone. Packing up after our time there and at the same time looking for fingerprints on the toys of our presumably dead son..." I cannot bring myself to look at Bertil. I know how difficult it was for him to be left behind after we had gone back to Sweden. "You did an incredible job Bertil. It is always difficult to identify such a young person. His teeth were perfect. We've been able to identify others from all their visits to the dentist, but with children…"

"We had so much contact back there Jonas, and you were such a support in the midst of it all. I'm glad that Ann and I met you in Khao Lak in February." *Imagine that he's been there all the time*, I say to myself.

I shut out Jonas' voice, take my eyes off him and begin to stare at the fridge behind him. I see Gösta lying on a stretcher, as if on a conveyor belt, surrounded by other bodies. I see the identification staff performing a clinical examination, clothed in protective gloves, mouth-guards, green overalls and boots. What position is he lying in? Is it the same position he was in when he died? Is he still stiff with rigor mortis or has the heat made his body limp? Or has his body rotted so much that it barely holds together? I do not seem to be able to stop torturing myself with these questions. Gösta, who was always so attentive to his appearance, who always made sure he looked presentable. Was he ugly now? Would he find it shameful to be lying on a stretcher with a swollen face and his tongue lolling out of his mouth, all blue-black? Are his eyes closed or are they staring into eternity? No, Gösta! I see that you're simply asleep and your skin is pale and smooth and it's only your dusting of freckles that are changing tone as they shimmer faintly brown. Your eyes are closed and you are dreaming wonderful dreams. Good night Gösta.

Two days earlier, on April 26, I had attended a memorial in Kungsträdgården in Stockholm. It was a beautiful spring day, the cherry trees were in full bloom and the sun was slowly sinking against a clear blue sky. A saxophonist was playing gentle music and the edge of the fountain was decked with five hundred and forty-three candles – one for each Swedish citizen whose life had been extinguished by the great wave.

Two years earlier, we had decided to exchange our lives in the countryside near Enköping, with goats and chickens, for a life in the pulsating metropolis of Singapore. It had been a time of anticipation, but also nostalgia. We had decided to leave the tranquillity of rural life and head for a multicultural city on the other side of the earth.

We had sold our house and I was in the process of sorting and packing up our things and trying to introduce Julius and Gösta to some English. Bertil had already gone to Singapore and was mailing every day with suggestions for places we could live. It was always so thrilling when he sent a picture and description of a possible home for us.

Ärna Airport

Gösta arrives at Ärna airport. On May 3, a sullen grey day, we are to greet him in his coffin.

I drive my Toyota from Stockholm with Julius and Karin in the car. This car has taken on a special meaning because it was a car that Gösta had travelled in. I had tried to sell it before we left for Singapore. It was from 1995, the year that Gösta was born. But it had proven difficult to sell for a decent price so I kept it and left it with good friends in our village of Grillby.

I do not have so many everyday things left that remind me of Gösta. His bedroom is no more and his school is just a memory. Everything is gone or is new. But not my Toyota. The car is still here and is one of the few reminders I have of the time when Gösta was with us.

We are driving along the motorway to Uppsala as if it was any normal day. I am driving, just as I used to, with all my children in the backseat and we are passing IKEA on the left. The children and I have been inside that IKEA so many times to pick up the various things a family of our size required. All of these memories are so familiar and obvious. They seem neither melancholic nor distant. It feels completely natural to continue along the motorway towards Uppsala. Up ahead we can make out Uppsala cathedral in the middle of the city where Julius, Gösta and, later, Karin too had run around demanding ice-creams in the middle of the summer. Is that now or was that in the past?

We are getting closer to the airport.

It is not until we take the exit signposted to Ärna airport that it hits me like bolt through my entire body. It is only then that I truly realize that we are on our way and why. We are to be reunited with Gösta after four months and one week. Four months and one week of not knowing where he was. Now we would finally be able to get close to him again, though he is indescribably far away. We will not be able to see him, but only the coffin that he is lying in.

My little boy Gösta, who was once inside my belly, who I nursed at my breast, who called me Mamma, whom I watched as he grew from an infant into a boy; now I understood that he would be lying in a coffin, dead.

I knew the coffin would be draped with the Swedish flag, I knew how the ceremony had been performed in Thailand. I knew what to expect.

As we approached the parking area the tears begin to block my vision and in confusion I put on the windshield wipers. But I manage to navigate my way around the parked cars and I switch off the engine. Julius and Karin have already got out of the car by the time I manage to pull myself together and get out.

Small groups of people are standing around the parking area, silent, as though they are waiting for something. I glance around until I see Julius and Karin, who have found Bertil and his parents and sister with her family. They are together with my father, my sister My, and her family. I walk wordlessly towards them.

We stand motionless for a while until the other groups slowly begin to break away and walk towards the large building some way up the rise. We trickle into the building.

Inside, tables have been laid with coffee and cinnamon buns. There are many groups of tables and on each table is a sign. The people begin to walk towards them and search the signs. Today, there are seventeen signs, each bearing a name. It might just as well have been the name of the person who had ordered coffee, but it is not. We wander about, looking for a sign with the name Gösta Schmidt on it. Today, there are seven signs bearing the names of children.

After a while, the priest begins to speak. He welcomes us and says how sorry he is for the reason we are gathered. He tells us his name and explains what is going to take place when the coffins arrive. Then police will wander between the tables to answer questions.

When he has finished speaking, everyone returns to their coffee cups and buns. I continue gazing at him; he looks so peculiarly normal.

I continue gazing at all the heads, bowed over their coffee. This too looks entirely normal. How is this possible? Maybe I am wrong. Maybe this is normal and it is just me who has bizarre thoughts...thoughts about everything being surreal. How can we possibly be behaving so normally? We are sitting drinking coffee at tables bearing the names of our dead. Why are we not moaning? Why are we not crying out loud? Why am I not fighting this, refusing to accept it? Why are we so composed?

A policeman passes close to our table and I simply and obediently raise my hand to get his attention. I have to know whether anyone here in Sweden has looked at Gösta, if we can be sure that it really is him inside the coffin. The policeman reassures me that it is him. A pathologist has checked to be absolutely sure that it is the right person inside the coffin.

I pose a few more questions, questions that I already know the answer to or that are irrelevant. I just want to keep him there. It is as though he is a link, however tenuous, between me and Gösta and I cannot let it go. But eventually I cannot find any more questions and I am forced to release him.

After a while, some people begin to ready themselves to leave the room. More and more people begin to go outside and gather on the courtyard outside the building. The priest takes the lead and we slowly fall in behind him. We know what is coming and we can soon see the cars up ahead of us.

The cloud cover is heavy and the wind is stabbing at us with tiny, sharp raindrops. The people are bracing themselves against the wind and against what lies ahead. Their heads are bowed towards the ground. Some are gripping their collars, as if for support.

We are in the middle of the crowd and we are moving gradually towards the place at which we are to meet our loved ones. Seventeen cars are standing in a fan shape, facing outwards. Beside each car is a person dressed in livery, standing to attention and staring out into nothingness.

The group of mourners stops so that the cars are now in front of us. At the command of an officer all of those standing beside the cars turns towards the trunk of the car. They open it, pulls the coffin outwards and then salute. Another priest has come and he performs the ritual by reading a prayer and a short text. Then follows a minute of silence.

Gösta is in car number seven.

We take a few faltering steps towards the car. The coffin is covered with the Swedish flag and on it lays a single orchid. After four months, we are about to be reunited as a family.

Four months and seven days have passed since we were close to Gösta. Now, the family is together again. He is with us again, and yet so interminably far away.

We stand for a long, long time beside the coffin.

Instead of weeping, I try to comprehend the fact that we are once again united and that we will never again be together.

It is so hard to leave the coffin. Like Siddhartha, who waited a whole night for his father's blessing before he continued on his way to seek, I too could have stood by the coffin and waited. Waited, waited for ...

Siddhartha sought something beyond his reality. He felt that something was missing and he waited until he was permitted to seek it. He waited for his father's permission before he left and the only alternative was to wait.

This is how I feel as I stand beside Gösta's coffin. I wanted to come close to Gösta and the only alternative was to wait.

*****

I had decided where Gösta would be buried while we were still in the jungle. That same night, soon after the wave destroyed so much for so many people, I sat numbly staring into the blackness. I could already see how we would bury Gösta in Villberga cemetery, close to the house we had left a year and a half earlier. It was a cold, raw January and the snow lay heavily on the fields. Towards the horizon leaden clouds were melting into the dense, dark pine forest.

It is cold on the day of Gösta's funeral as well. 21 May 2005. Gösta would have been ten years, three months and one day old. There is no snow though and the trees are beginning to show a hint of green. But in the absence of sunlight, the land seems dim and sober.

I had had daily contact with the funeral director. There were so many decisions to make – everything from which flowers should decorate the coffin to what should be put to eat on the sandwiches. Sometimes I found it difficult to keep things clear in my mind and I would begin imagining I was preparing for Gösta's high-school graduation or his wedding, not his funeral.

"How many people have said they're coming?" I asked.

"Last time I looked it was ninety-eight."

"Goodness."

"But we can expect more because the 21st is still some time away. I think it'd be a good idea to have various kinds of sandwiches, with meat, fish or vegetarian fillings. That way we're covered so no one will feel left out."

"OK, I trust your judgment."

"And I think we should include light beer or mineral water, and sodas for the kids."

"Are their many children coming?"

"Yes, there are. But some of the parents are hesitant about bringing their children along. They're worried that it's going to be too much for them."

"I can understand that. But I think it's important that they come too, at least Gösta's friends. It won't be any less true just because they don't come."

"That's exactly what I said. I've explained that it can actually be important to come, to say good-bye, and it may not be as dreadful as one imagines."

"I want the children to come. They can't let Gösta down now."

"No. But alright then, we'll organize the food as we've discussed. I'll ring again nearer the time. Bye for now."

After putting the phone down, I gaze out at the lawn and the bushes through the window of our newly renovated house that we have already moved into. The sun is shining and buds are beginning to swell. Should I not be standing here organizing the catering for a big party? Why did it turn so wrong?

On the days leading up to the funeral, we meet a priest called Lars Collmar. He is going to officiate at Gösta's burial. He was the priest who christened Julius. He is calm and secure; I remember thinking that he seemed so wise. Those of us who are left in the family sit at a table together with him and run through what will happen at the ceremony. Both Bertil and I want to say a few words during the funeral and we each get to read out what we want to say.

When I read what I have written, Bertil catches my eye and nods towards Julius. The tears are streaming down his cheeks. I take his hand and continue reading.

Saturday May 21: The Funeral

We are the first to arrive at the church on this bitter day, Bertil, Julius, Karin and myself. On this cold, miserable day I shall not shed a single tear. I shall keep a tight hold on myself, I will not sway an inch. I will not feel anything.

Inside the church, we saw the white coffin in front of the altar. I nodded an acknowledgement and then went out.

It was beautifully decorated with the flowers we had agreed on a few days earlier with the funeral director.

We stand together for a while on the church steps and watch the first cars arriving at the parking area just below the church. Bertil's mobile beeps. Jonas has sent a message to say that he is with us in his thoughts on this day.

I can see Gun arriving at the church gate with her family – David, Gösta's best friend, and David's brother Markus with their father Pelle.

The last time I saw Gun was almost a year ago. I had had Gösta at my side, full of anticipation, his hand locked in mine. We had walked down Nybrogatan towards the Royal Theatre where I would leave him to catch the 47 bus. Gun and David would be on the bus to meet him, and then they were going to go to Gröna Lund amusement park. Gösta had been looking forward to this so much! For weeks, long before we had left Singapore to return to Sweden for the summer holidays, Gösta had been longing to see his friend David. And now that time had come. After a short wait, the bus arrived. I gave Gösta his bag and kissed him good-bye. He had hopped up onto the bus and I could see on his face through the window the joy that his friendship gave him.

Gun comes towards me, her face swollen from crying and I decide on my strategy: don't speak, just nod.

For the entire time that I go around greeting people I do not utter a word. I nod and keep control of every facial muscle. I allow all the tears and desperate words bounce off me. I am of steel.

As the church bells toll, I look around for Bertil, Julius and Karin. We walk together along the aisle towards the coffin where Gösta lies and we seat ourselves on the pew to the right.

The Speech I Gave at Gösta's Funeral

This is so hard. I hate having to see everyone gathering here. It is so hard having to be here myself with Bertil, Julius and Karin. Right now, we should be sitting with the children on a beach in Singapore eating ice-cream and drinking beer. Gösta would be asking how many days were left until we were coming back to Sweden. He longed for Sweden.

But because we happened to be at the wrong place, at the wrong time we are forced to gather here today and for that reason I am grateful and touched that so many of Gösta's friends are with us. Gösta is snuggled up in his own heaven.

The day before Gösta disappeared, 25 December, I told Julius and Gösta that they had been mentioned in Enköping's local newspaper and that there was a picture of them. Gösta was so proud of this. I think it was important to Gösta to be remembered so please, all of you, never forget Gösta. I hope you will all bear Gösta with you in your thoughts both now and always. Let Gösta remain in your hearts.

And to all the children who are here, please take him with you in your games when you laugh and have fun.

Lötgården

After the funeral the guests continue the six kilometers to the reception at the community center. When we arrive we see that the flag has been hoisted to full mast. Long tables have been covered with white tablecloths and there are flowers and large platters filled with the three kinds of sandwiches that we ordered.

Outside the room, Bertil and I have prepared a table presenting Gösta's life story. We have gathered almost all the photos we have of him from his babyhood up to the last pictures we have of him.

We have set up a computer with a slideshow of our final days with Gösta and it finishes with pictures of the place where the tsunami has swept over the land and destroyed everything. We had also made a video film showing our last Christmas with him, with Julius and Gösta opening their gifts and finding mobile phones inside and Karin discovering her gift of a white toy cat that can sing. The last thing you see is the wave far out at sea. Then the screen goes black.

A small crowd gathers in front of the table while Bertil and I stand to one side, watching.

"How many parties have we been to here at Lötgården?" I ask rhetorically as I turn towards Bertil.

"Many. One of the first we planned to have was Gösta's christening."

"And in the end we held it in the little cottage. But Gösta must have taken some of his first steps here on this floor. His and Julius' birthday parties, new year's parties with all the children, Christmas markets ..."

I look out through the window and see children running around outside. They do not seem bothered by the fact that it has started to drizzle. I cannot bear to think of the fact that Gösta is not among them. But then I can see him running past the window with all the other children, he is there, together with all his friends, right where he belongs and where he is happiest.

"And now we've organised one more party for Gösta and all the children."

I continue watching the children, and I smile.

Timeless Ponderings

I think of the last time I saw Gösta ... I think of his final breath. What actually happened when he died? When he took his last breath – what was the last thing he thought? What was he feeling? Was he feeling anything at all?

If he was feeling anything was it fear, anxiety, pain, panic? Did he feel let down, where were his Mum and Dad? Why were we not there to rescue him? Or did it all go so fast that he had no time to think? Did something hit him on the head so that he lost consciousness, or did he get trapped under the water and struggle to get to the surface, completely conscious? Did he have time to breathe the dirty salt water into his lungs? Did he have time to feel panicked when he pulled water instead of air into his lungs? Was he relieved when he finally took the step into death? Did the panic stop then? I have always heard that drowning is a terrible way to die. What does it feel like to die?

Gösta knows, but I do not. And he will never be able to tell me what it is like to die.

And I let him down. At the moment that he needed me more than ever, I was not there for him. I shall have to live with that knowledge for the rest of my life.

A Desperate Thought About Hopelessness Summer 2005

The thing I have fought for is ripped from me.

Life together with Gösta feels like a loop. A journey I had believed was my life, but which turned out to be just a dream. It was never supposed to be my reality.

Ever since I was a child I wanted to have many children – four or five perhaps – but three was just fine. Soon after Julius was born I began longing for Gösta. When Julius was nine months old I found myself wandering through the medical fair in Älvsjö looking for pregnancy tests. I still had not begun menstruating again after Julius but I still hoped I was pregnant with Gösta. I wanted Julius to have a friend as soon as possible, and he arrived when Julius was just two years old.

I was so proud of my sons.

Julius and I were such a team when Gösta was born; it was he and I together who took care of Gösta. It was we two who checked that he was alright, that he was dry and fed and comfortable. When I breast fed Gösta, Julius used to pull up his t-shirt and feed his doll. And when Gösta had fallen asleep we would curl up on the sofa and read a book together. Julius was enthralled with his little brother and when Gösta was a little bigger, he began looking up to him. They soon became the best of friends.

Even before Gösta was born I could feel that I had a Gösta inside me. He would be called Gösta after my grandfather, whom I never met. Julius was named after

Bertil's great grandfather, but Gösta would be named after one of my forefathers. Gösta is such a finely rounded name.

When Gösta was three, I began longing for another child and when Karin finally arrived a few years later I felt so privileged to have my three children.

I do not believe in destiny but in my darker moments it feels as though some greater power had decided that I should not have three children. It was as though I had battled against destiny and won when Karin was born but in the end destiny took revenge and stole Gösta from me. I was never meant to have three children.

Was it worth it? Was it worth doing this loop? Is the happiness I felt for the nine years I had with Gösta worth the agony of losing him? Would it have been better if he had never been born?

Is it worse never to have seen than to become blind after seeing?

July 26: Placing the Urn

It is overcast again on the day that Bertil, his sister Eva and I go to Villberga church to place Gösta's urn. I take my own car and I arrive at the church a little while before Bertil and Eva. Those minutes give me a chance to take a few deep breaths and clear myself of thoughts. I have time to look out through the window and see that the sky is grey and the landscape is colourless before Bertil pulls up beside me and opens his door.

We walk together towards the church door where we are met by a caretaker who leads us into the church. I do not know what I am expecting but suddenly we are inside the little church with the pews on each side of us. And just before the altar is a table covered with a long, white cloth that reaches almost to the floor. There are flowers on the table and in the middle, an urn. It looks like a Greek vase, clean and white.

The three of us draw together and put our arms around one another as though we are one being. Tears well up from somewhere deep in our stomachs but from inside me I can also feel laughter bubbling up. I cannot control it. I laugh at the same time as I cry. I am standing here looking at a table with a white cloth on it and on the cloth stands an urn and in the urn are the remains of Gösta – the remains of Gösta.

I am unable to do anything about my laughter. Gösta has sat in this church with his school friends and sung Christmas hymns before the winter holiday and summer songs before the summer holidays. He stood up by the altar as a proud big brother in this quaint church when Karin was christened. And now his remains stand here in an urn on a table with a white cloth on it.

I laugh.

Bertil pulls away and walks up to the table. He takes the urn between both hands and hugs it to him.

"This urn weighs almost as much as Gösta did when he was born," he says as he cradles the urn like an infant.

Then he walks past us and continues down the aisle and out of the church. Eva and I turn and follow him.

The caretaker, who has kept his distance, now advances towards us to show us the way to Gösta's grave.

When we reach the grave Bertil sinks down onto his knees so that he can place the urn in the ground. A hole has been dug just the right size to hold the urn. He looks at me and wonders if I would like to hold the urn or maybe put it in the ground myself. But I shake my head. I do not want to feel the weight of Gösta as he was when he was born.

Early August

Bertil knew that we had a video film somewhere with Julius and Gösta training tae kwon do and some other clips from Thailand and Singapore. After searching high and low we finally come across it and then we sit ourselves down expectantly in front of the screen to watch it. It is still summer outside and even though the sun is making its descent now, the August heat is wafting in through the open French windows. Julius is sitting on the floor at my feet and Karin has crept up onto my lap. Bertil is to the side of me in an armchair holding the remote control for the video. He presses the start button.

I feel a strange flicker of excitement as the film starts, partly because I do not know exactly what is on the film and am awaiting the unknown, but above all because we are about to see Gösta.

The video contains some surprises. We see Gösta and Karin playing by the pool in Singapore. There is Gösta helping Karin turn on the shower and then they are jumping into the water. Julius and Gösta are training tae kwon do – Gösta looks uncertain of how to do the moves and he is stealing a look at Julius, who is performing his with confidence. Then we see the children jumping about on the rocks by a waterfall surrounded by Thai children out in the jungle not far from Chiang Mai. And now, it is Karin who is filming and the image leaps about showing sometimes only the treetops.

I sit huddled up on the sofa, relishing these visions of Gösta on the TV. He is really there. I feel how my face softens until I find myself smiling. He is so close. I can see him moving, laughing, talking, all so naturally. I want to see more and more – my thirst to see him is unquenchable. The more I see, the greedier I become for more. I almost forget to breathe and simply allow myself to be filled with the moving images on the screen before me. Everything I see looks so natural and I am captivated by the memories of a time I know so well. Each sequence shows new, but familiar experiences and I cannot get enough of it.

And then it is over. Gösta dissolves with the film and I am left sitting with an idiotic grin on my lips. The sweetness that spread through me while the film was running remains for a few moments, but only a few moments. Slowly at first, but then with gathering speed, the sweetness ebbs and leaves a bitter taste behind. In my thirst, I had gulped and gulped as though it would never end only to find that I had been drinking sand; it was all simply a mirage, an illusion.

My throat tightens, trapping the air and preventing it from reaching my lungs. My grin twists and stiffens as emptiness grips me.

No one says a word. We are left silently staring at the dead screen, but after a time we begin to break away. Julius gets up and I hear his footfall on the stairs as he goes up to his room to start up a TV game. Karin disentangles herself from my arms, slips down to the floor and then disappears to her room too. Bertil and I sit motionless. Perhaps he too is wondering what the next step will be.

Autumnal Chill

I am imprisoned for life. I shall never be able to leave my cell – a cell that is ugly, bare, colourless and chilly. In the space of a split second I was mercilessly thrown into this cell on 26 December. I had no idea why.

At first, I refused to accept that I was here. I just sat there, apathetic on a rough, brown grey horsehair mattress that lay on a rusty iron bedstead and I stared ahead of me. Inside, I was planning my escape because there was no way I could stay here.

My strategy was to behave well because then my prison guard would be bound to realise that a mistake had been made. I was pleasant and friendly.

When this proved fruitless I began to scream like a child so that the guard would take pity on me and release me out of sympathy. But even this did not help.

Okay, I will just run away, I thought, but I soon discovered that this cell is a place one can never escape. I gradually came to realise that there is no way out. Well, there is one way, and that is to die oneself, but this was no alternative.

Now I am here in my cell and must understand the truth about my life and learn how to relate to it.

I have decided not to die. This is the first choice, and then there are other choices that have to be made. How do I want life to be inside this hideous cell? Either I can sit here and allow it to continue to be hideous, cold and bare or I can begin to furnish it. I can try to make it as attractive as possible in here now that I understand that my guard will never release me.

And so I have begun to decorate and my prison guard has become more amiable now that I have begun to accept the fact that I have to stay here. I am allowed to receive some of the things I need. I have requested a softer mattress and I have repainted the bedstead. I have thrown a colorful bedspread over it and soon I shall be getting a TV.

I will never feel comfortable inside this cell, never. But I shall be able to make it liveable, for the time that remains.

Think of Julius and Karin.

9 Months and 7 Days

It feels important to analyse my thoughts and organise them so that I can keep them under control. They shift from day to night. They are easier to curb during the day and I feel able to steer them and keep myself in the here and now, observing what is going on around me and allowing the everyday to roll on.

At night, it is much harder – time and space dissolve. Maybe it is because I am lying down that I lose the power of control? The darkness does not help – lying down in a dark room with one's eyes closed. Then I am sucked out into the universe, observing how the Earth recedes and feeling how time becomes inconsequential. It is then that my longing for Gösta becomes so heavy. I lose all points of reference and their interconnection and context. A vacuum expands. My presence on Earth is so meaningless; death beckons and I long for my time here to be over. My allotted period of existence is a prison and I want to break loose. My time on Earth is simply a long wait, a passage onwards.

Life is something I must endure. I have a bird's-eye view on it. There is no point in trying to flee too early and I shall never do that. I accept my destiny. But it is like holding one's breath for a time. Like a long car journey. I know roughly how long the journey should take and I know I have to see it through because there is no alternative. I know I shall arrive, but I need patience to wait it out. While I wait, I may as well make the best of the situation and arm myself with sweets to suck on, good music and maybe pleasant company. In daytime, it is easier to be present in life, on this long journey. Here and now.

Think of the little things and avoid looking upwards because it is so hard to look at the heavens.

At night I am removed from all this and am drawn into oneness. The most difficult things are falling asleep and waking up, in the borderlands between sleeping and wakefulness ... or maybe when I am trying to get to sleep, lying there still with my thoughts floating freely. Then it is difficult to rein in my thoughts by keeping busy with everyday tasks. It is also hard in the morning, when I am wrenched from sleep and have to gather my thoughts for a new day.

I think again of Siddhartha.

"When a person is asleep, in the deepest sleep, she is in her innermost self and is at one with Atman."

Atman- the deepest part of every being. The indestructible innermost part of a being is Atman, the absolute. It does not exist in time or space, but is free and blissful. It is the greatest and smallest aspect of everything. Everything, in its entirety, may be found within it.

I have decided not to be miserable when I dream about Gösta because this is when I am able to meet him. Maybe it is in my deepest dreams that I make contact with my soul, and thereby with the universal soul, and maybe this is connected to Gösta's soul and it is this that makes contact possible.

In any case, I delight in having dreamt about Gösta and that is why it is so hard to wake up and be thrust into daily life again.

9 Months and 8 Days

How would I have reacted if I had known twenty years ago that my nine-year-old son would die? I have always wanted to be surrounded by children and assumed that I would be. If I think of it like this, it is intolerable.

Losing a child is the worst thing that can happen and it has happened. So what is left?

How is it possible for me to continue thinking when the worst possible thing has happened? How come my liver, my kidneys, my stomach continue to function? How come my body continues to function as though nothing had happened? Why does my heart continue to beat and I keep on living? It is strange.

Maybe it is because I had to decide from the start whether or not to continue living. I had to decide. Do I or do I not want to live?

I want to.

I think of Julius and Karin and I know why.

Pretending

It is crazy but reality seems like a bad film.

I'm sitting on the bus that has just stopped at Nybrokajen to let some people off. The bus in front of us has also stopped and a group of children of about eight or nine years are jumping off with their teacher. And among them I can see Gösta together with a friend. He has got his jeans and t-shirt on as usual. His hair – I can tell that it is him by the hair – is a bit long and unkempt. It looks as soft as usual and it gleams blond with shades of copper caught by the sun. He is walking now with his friends away from where I am seated and he disappears into the crowd. I feel a familiar sting at the back of my throat but I manage to calm it.

The bus continues on its way and I get off at the next stop. I start walking towards Birger Jarlsgatan, towards the Royal Theatre. The weather is magnificent and everyone is out soaking up the last warmth of the summer. As I pass the steps leading up to the Royal Theatre I see him again. He is standing there together with his classmates to eat some fruit. I stop. He is just a few meters in front of me with his back towards me. I stand motionless, watching how he eats and how his hair is ruffled by the wind.

Suddenly I feel that sting in my throat again and a few seconds later the tears are coursing down my cheeks as the spell breaks. I am thrown back into reality, or was it back into that awful film? I step out of myself and observe myself standing there, staring at a young boy as the tears stream down my face. It is so theatrical and pathetic! I chide myself for behaving like a second-rate actress in a badly-made film. "The devastated mother stands alone, weeping as she watches a child in the park as she thinks of the child she once had, but who is now gone forever ... blah, blah, blah ..."

Why do I put myself through this? When I first noticed the children getting off the bus why did I try to pick out one that looked like Gösta? And when I then find one, why do I carry on torturing myself by pretending it is him? Why don't I just look the other way and keep myself anchored to the moment, get a handle on my thoughts and focus on the "here and now?"

Maybe it is because it gives me pleasure, even if only for a short time, to pretend that everything is normal and for a few minutes feel that I'm like everyone else who is out there enjoying the September sunshine without a care in the world. Just then, I don't care if the bubble is going to burst any minute, when my body will no longer permit me to deceive it and the stinging sensation again warns of tears.

Why do we no longer use mourning bands?

November – December

I am so envious of a woman I met about a year ago. She told me she lived overseas with her family – somewhere in Southeast Asia I think. She said she had three children: two boys of nine and eleven and a little girl of four. They had lived overseas for more than a year and were enjoying it.

She was also a little anxious about her children because they had been so happy where they lived before, in Sweden. She asked herself whether she had done the right thing to uproot them all and take them away from their secure environment to a place on the other side of the world. But she and her husband had agreed that in the long term the children would gain so much from having lived abroad. They would learn English and experience so much that would enrich them for the rest of their lives.

She told me that they had taken the opportunity to travel widely in the region. Their friends at home thought it was so brave of them to travel like that with three children but she said they had only had positive experiences.

However, she also told me that the children were homesick, particularly the middle child who was nine. He wanted to move back to Sweden because he missed his friends so badly. This must have been just before Christmas because I remember her telling me that she was concerned that the children would have preferred to spend Christmas at home in Sweden. Instead, they decided that after all their adventurous trips they would have a relaxed and safe holiday in Thailand. I heard from her briefly ... it must have been on Boxing Day last year ... but since then I've heard nothing. I don't know what became of her.

I've often thought about her but I'm afraid that she now belongs to another world – a world that is inaccessible to me and for this reason I envy her so deeply. I know that I shall never again experience what she described about her children.

The First Christmas

And soon it is Christmas again.

There have been discussions about how and where to celebrate Christmas, but inside I just ask myself why it should be celebrated at all. I've been searching for the fast-forward button but it is futile. Instead, I'm sitting in the Toyota with my father, Julius and Karin and we are on our way to Falun, where Bertil has already arrived at his parents' home.

It is 22 December 2005, the darkest and shortest day of the year. Darkness is gathering outside though it is only three in the afternoon. A fog has settled over the road. We have just left Täby and we are skidding onto the road for Sollentuna. Slush slaps against the side of the car like waves against a speedboat. Oncoming cars are also kicking up slush onto my windshield and for an instant I can't see anything. The window wipers beat frantically and I keep pressing the button for the washer fluid but it doesn't help. The windshield simply becomes more and more caked until the washer fluid is finished and I have to turn off in Sollentuna to find a gas station.

When I open my door, the cold air bites and I feel the ground squelch under my feet. After a couple of steps the wet snow is already seeping through my shoes and making my socks wet. I shudder and hurry towards the shop. A few minutes later we are all sitting in the car consoling ourselves with sweets. The warmth spreads quickly down to my wet feet and a distinct smell wafts past my nostrils as I push the washer fluid button. The visibility outside is far from good but at least I can see through the windshield now.

It is soon quiet in the car because the others have fallen asleep. As we approach Grillby, I begin asking myself whether we should stop to visit the grave or not. I don't come to Grillby all that often so when I do I always visit the grave. But with a stab of guilt I decide that today we will not go there but instead just drive on. After a short while the exit sign appears and instead of turning off to Grillby, I continue strait on. Gösta is buried just a kilometer away.

It is the 22 of December and it is nearly four o'clock in the afternoon, so it would be around ten o'clock at night in Thailand. Exactly one year ago we were sitting at an outdoor restaurant finishing off our dinner and sipping exotic fruit drinks. Now I remember, that was the day that we had been taking the motorboat out to snorkel on the Similan Islands.

It had been a beautiful day, snorkelling over the coral, with white sands, turquoise water and rich greenery and we had seen so many colourful fish. On the way back to the mainland we all had sea legs after the one-hour motorboat trip back and Julius and Gösta were glowing with sunburn because they had been sitting at the front of the boat in full sun. What irresponsible parents we had been!

The silence and darkness begins to break up around me. The monotony of the road gives way to the glimmer of lights and the sound of the sea. Although I am so focused on the blinding lights of the oncoming traffic, I begin to relax and can feel my face smoothen as I continue to remember from one year ago.

My passengers sleep almost the whole way up to Dalarna but as we near Hedermora they start to wake up.

"When you see a sign to Falun you think you're almost there but you're not. The last bit is actually the longest."

Julius has just woken up and he is looking sleepily out at the darkness.

"That's not so strange! You've been asleep the whole way," I say.

"You looked so comfy with your head snuggled up against that pillow you'd pushed up against the window," says my father from the back seat.

"Agh, I've got a sore neck," says Julius, rubbing at the muscles with both hands.

"You snored!" says Karin from the other end of the back seat. "I haven't slept at all and now I need to go to the toilet."

"Ok, I'll speed up, I say, and I put my foot harder on the gas now that the traffic has eased."

It feels so good when we finally arrive. The house welcomes us with all its lights. The verandah is decorated with spruce and there are two beautiful brass tubs of it on the steps. Indoors, Bertil greets us and we can smell the spicy aroma of mulled wine.

It is only now that I realise how tired I am. Despite all my memories in the car, I have been tense. Not only my eyes, but also my shoulders and back feel heavy.

Bertil takes us down to the basement and as we go we can hear the crackle of an open fire down there. That is where the mulled wine is. On the table in front of the fire there is a large jug on a stand, warming over a tea light. Little green cups are lined up and there are bowls of raisins and nuts. And there is a larger bowl full of M & M's and a red basket with gingerbread cookies. The smells mix with those of the fire and the scent is warm. We sit ourselves down and Bertil gives us each a cup.

"Hmm, it'll be good with a cup of mulled wine."
My father is leaning back comfortably in the sofa.

"Yes," says Bertil, "even though it's so sweet it's still a special flavour. I think that's because we associated it with a particular atmosphere. Darkness, an open fire, warm inside and cold out and Christmas approaching. For me,

Christmas means peacefulness. At least I'd like it to be like that."

"Hm, I have to hang onto the moment," I add.

This Christmas, that I've been so dreading, actually feels warm and inviting and I must accept it. But I can't avoid thinking about the last time we drank mulled wine.

"Do you remember, Bertil? You remember when we invited Lissy and Joseph over for mulled wine a year ago?"

They were from Singapore and had never tasted it but they thought it was an interesting experience – they liked it. The children were so excited about showing off our Swedish tradition.

"Yes I remember, and I remember how we managed to make it feel Christmassy despite the thirty-degree heat."

"I remember that too," says Julius, who is sitting on the floor in front of the fireplace with his mouth full of gingerbread. "We'd bought masses of sweets and chocolate and were really surprised that Lissy didn't think our sweets were sweet!"

"Yeah but you know what a sweet tooth the Singaporeans have got. At work there were people who would add sugar to their Coca-Cola," says Bertil.

"I wonder how Lissy and Joseph are getting on now, after their move to California," I muse. "At least they don't have to keep their relationship a secret like they did in Singapore. Imagine, in this day and age that a Chinese person can't have a relationship with an Indian. Even their parents wouldn't accept it."

I lean back in the sofa and ponder over my memories from last Christmas. I think of the dance Julius and Gösta made up and performed for us to the accompaniment of George Michael. Of course they received a standing ovation. I smile.

Almost unwillingly and despite my bittersweet memories, I allow myself to be absorbed by this easy atmosphere. I try to command my thoughts and remind myself of the "here and now." I rest my head on the edge of the sofa and close my eyes.

My father and Bertil continue chatting while Julius and Karin gobble sweets and mandarins.

And then I feel how the distance grows between me and all that is around me. I can't tell whether it is me who is moving away from all the sounds around me or whether it is the sounds that are retreating. It makes no difference; it is just good to feel this distance increasing as the wine begins to work inside me.

The next morning the sun is shining and the snow on the lawn is sparkling. When I come down to the kitchen I find that Bertil has already given Julius and Göst ..., no, I mean Karin their Christmas rice porridge. For a split second, I was back in a previous life before being hauled back into this one just as quickly.

"Good morning," I said.

Julius and Karin both turn around, cinnamon dust all around their mouths.

"Hi," they say.

"Urban and Eva will arrive soon. They rang earlier," says Karin excitedly. "It's going to be such fun to see Thomas, Amanda and Johanna."

"Amanda and Johanna," says Julius. "They're like eight and twelve years older than you! They won't want to play with you."

"I hope all you kids will have fun together," Bertil says as he ladles more porridge into the children's bowls.

"I don't want any more," shouts Karin as she jumps up from her chair and races off.

"And Auntie Emy will be here with Sebastian this afternoon."

Julius finishes his porridge and then he takes off as well.

"Gee, what a gang," I mutter with resignation as I pick up Karin's bowl. "I feel so divided. On the one hand it is so lovely to spend Christmas with all the relatives in a roomy house with space for everyone. It is wonderful to see all the cousins enjoying each other. That is just the kind of Christmas I always wanted for my children. It was just such a Christmas that I promised Gösta a year ago. And that is exactly why it feels so dreadful. But I have to hold together for the sake of my children."

Bertil has sat down at the table with the porridge pan in front of him.

"Yes," he says with force. "You can almost hear Gösta's happy shouts. It is as though they're in the walls."

"So I guess we just have to get on with it." I get up and pour myself a cup of tea.

After clearing up the breakfast things we gather everyone and go for a walk. When we return, Eva has already arrived with her family. The house key was where it always was so they let themselves in. And that is the end of the peace. The children start to play together and they are tearing around the house while we adults are working away in the kitchen. The afternoon and evening are so busy that I hardly have time to think of anything other than tasting all the delicacies and keeping an eye on the Jansson's temptation that is bubbling in the oven.

That night I fall asleep without first tossing and turning and it feels good.

Our Christmas looks just like any other Swedish Christmas. If you were to look through the window you would see three generations of a family together. You would see excited children, all dressed up and eager,

glancing up at the window without meeting your eyes –
they would be looking for the arrival of Father Christmas.
You would see all the food and drinks, the chaos and
laughter around the table and how it would suddenly still
as the clock struck three, when Donald Duck's traditional
hour-long program on TV starts. You would turn and
leave with warmth in your heart, thinking that this must be
what a happy family looks like.

It is December 26.

At nine in the morning I get up and go to the kitchen
to make myself a cup of tea. Then I go upstairs, position
myself on the sofa and switch on the TV. I am prepared.

It is pouring with rain in Khao Lak. The ceremony has
just begun and everyone is sheltering under umbrellas and
parasols. The rain is symbolic. "What else could it do but
rain in Khao Lak just now," says the priest in his speech.

Tears begin to free me. At last I can let go. I can watch
the ceremony in Khao Lak all day on the TV and see my
friends from the tsunami being interviewed and I can feel
their presence.

Some hours later, darkness has embraced Khao Lak
and the people are preparing to send 5,000 rice balloons
into the air. Then I receive a text message. It is from Åsa,
saying she will send up a balloon for Gösta together with
the one she is releasing for her Amanda. It brings me still
closer to Khao Lak.

I hardly notice when the rest of the family comes and
goes during the day and I have not noticed that I am
sitting alone and that it is now dark around me except for
the flickering light from the TV. At some point during the
day someone brought me a plate of food and my teacup
has been refilled. Now the cup and plate are empty and I
assume it was me who ate and drank.

Ten hours pass before I get up. My legs are stiff and I stagger out to the bathroom. My cheeks are burning and my eyes are puffy. But I was prepared.

A year has passed.

***

"How was Christmas?"

"It was great thanks. Everyone was there and there was masses of good food and presents."

Or:

"It was ghastly. I thought of Gösta every second, thinking that now it was exactly a year since he did this or that. Now he has been dead for exactly one year."

PART 2
2004

Singapore, December 2004

I sit down on the bus heading into the Center. It has just stopped raining and heavy clouds are still hanging low in the sky – the air is warm and moist. Although I have just showered I am already sticky all over. It is virtually impossible to have light enough clothing in Singapore. My thin, sleeveless top is clinging to my sides and my shoulders are stuck onto the green chair back. A cool stream of air from the air conditioner hits me in the face and it feels good but if I let it carry on I know I will soon be chilly, so I reach up and point the vent away from me.

I love the double-deckers that still roam the city. I sit on the upper deck at the front and feel the thrill of a child who is allowed to sit in the front seat of the car or indeed at the front of the top deck of a bus. Looking at the traffic from above gives a new perspective and a new dimension to one's thoughts. It is easier to get an overview of one's thoughts and that is particularly important today. Our last Christmas in Southeast Asia is already approaching and we have been stressing this to the children. All three of them want to spend Christmas at home in Sweden with all the relatives, with snow and a Christmas tree and piles of gifts. We are trying to persuade them that after this Christmas, all Christmases will be spent at home so we really need to make the most of the tropical heat and spend a memorable Christmas here.

It is 10 December and Julius and Gösta are going to participate in the Swedish school's traditional St. Lucia celebrations this afternoon. Julius is so big now that he thinks it is a little embarrassing to dress up in his white costume but Gösta is looking forward to dressing up in his red and white one. This morning, before he left for school, he took my long-sleeved red pullover and the Christmas hat that Bertil had bought. It is bright red and the hat is shaped in a spiral so that it stands upright with a white pom-pom on the top. Gösta loves it and when he puts it on his whole head wobbles.

It is one o'clock and I am on my way to one of our favourite lunch cafes to meet up with Bertil. I am famished and am looking forward to some good food and a Tiger beer. Over lunch we are going to discuss Christmas and what presents we are going to get for the children. We have decided to spend Christmas in Thailand – we want to relax and spend a holiday that does not involve any challenges. It is to be our last Christmas overseas so we want to enjoy some snorkelling and delicious Thai food washed down with Singhas. Bertil and I are not too bothered about Christmas traditions but the most important thing is to make sure the children enjoy it. Christmas is for children.

Thailand will be something completely different from all the other trips we have made since we came to Singapore. I suggested we go to Khao Lak, which is about an hour by car from Phuket in southern Thailand. I have been there once before together with Gösta and Karin. We were there with my sister, My, and her two daughters, Olivia and Andrea. We discovered our own little paradise, a locally owned guesthouse by a magnificent beach with benches that were roughly hammered together and tables wedged between the pine trees. Gösta had made friends with a girl his age from northern Thailand, the daughter of the couple who owned the guesthouse. They spent half the year in Khao Lak and the other half in Chiang Mai, in the North. We had decided to get away from the popular tourist areas and were so pleased to have found this little oasis. The children ran about on the beach looking for shells, returning now and then to pig themselves on mango, Gösta's favourite fruit. After we left, I promised that we would return one day, and since Julius had not accompanied us, I wanted to show him where we had been.

The bus lurches to a halt. A cyclist turned just a little too late and escaped being run over by a hair's breadth. I am shaken out of my dream and I shift my weight on the seat. The bus trip is slowed by all the stops and traffic lights and I start regretting that I did not take a taxi instead.

Business is throbbing on the streets outside. The pedestrians squeeze their way between all the fruit and vegetable sellers that are interspersed between the food stands that are so typical of Singapore. I remember how I thought they looked so grubby and cheap when I first arrived in Singapore, with all those plastic chairs and tables. Quite unappealing. All those vats of food behind the counter – it looked like school food. But I soon found that they sold mouth-watering cuisine from all over Asia and that it was a delight to sit and eat at them.

I can see groups of schoolchildren sitting at the tables now drinking sodas. They are dressed in uniforms and have their exercise books in front of them. School in Singapore is tough. The country lacks natural resources and even has to import water. It is completely dependent on foreign investment and this makes for a highly competitive environment. Only the strongest are able to get ahead. The children are drilled in lectures, even at pre-school. My little Karin attends a Singaporean pre-school and is already studying maths and Chinese.

We are slowly approaching the Center, past Raffles Hotel where they serve the traditional, thirst-quenching

Singapore Sling. At the next stop, I get off the bus and start walking towards the restaurant.

Bertil is already there and has two beers in front of him. I sit down and wrap my hands around one of them. The short walk from the bus stop was enough for the heat to creep through my whole body and I gulp at the cool beer.

Over lunch, we talk about what to buy the children. We decide that we should limit Christmas presents since we will be travelling. Karin can have some little things, since she is so young but Julius and Gösta will get one fine present each – a mobile phone. We justify our decision by saying that it is helpful for us that they each have a phone. We have often wished they had one when they have been out visiting friends and we think they will be so pleased.

We spend the whole afternoon wandering from shop to shop comparing the prices of mobiles and subscriptions. We agree that it is best to get them prepaid phone cards because in only six months we will be returning to Sweden so we can buy them full subscriptions then. The phones we settle for are luxury models. For Karin, we pick out a toy cat that makes a meowing sound and a few other small items that we are convinced she will be pleased with. Then we throw ourselves into a taxi to go and collect Karin before we hurry off to the St. Lucia celebrations at the school.

How many times have I listened to the St. Lucia procession approaching from a corridor outside the room, listening as the children's singing grows louder? We are sitting, parents and siblings, twisting round to see when the children arrive, singing in harmonies more or less in tune, dressed up in their various costumes in reds and whites and golds. The Lusselelle song and all the other familiar songs blend with the rain that is beating against the windows. The room is filled with the smell of gingerbread and saffron buns. When the children have completed their repertoire, the procession trails out into the corridor again.

A while later, the children return, still in costume and they begin racing around. Karin has put on the brand new dress that she had changed into from her uniform in the taxi on the way here. Gösta and his friend have stocked up on saffron buns and hidden them under a bench, but they soon forget about them. The pleasure of running around in the excitement of the coming Christmas holidays eclipses the greed for buns. Julius is among the eldest of the children and he has already outgrown the racing around stage. He comes and sits with us instead and asks if we'll be going home after this or whether we're going out to eat somewhere.

Despite all this, it is difficult to take the celebrations seriously. The atmosphere refuses to be completely genuine as the monsoon outside thrashes palm leaves against the windowpanes and lightning ignites the sky.

When the children have begun to calm down we gather together and brave the rain outside, running towards a taxi that seems to be waiting just for us. We scramble into the car, drenched and laden with bags, but we are in high spirits. The taxi pulls away into the dense Friday traffic towards a restaurant by the beach.

I keep daily contact with the travel agency in the run-up to the end of term. Tickets are running out. But finally I receive news that the travel agent has managed to find us tickets to Phuket. The only trip available is one leaving the same day that the boys break up from school and returning on Christmas Eve. Typical, I lament, it'll be chaos trying to leave that day and it isn't ideal having to return on Christmas Eve either.

I discuss it with Bertil and we decide that it is better than nothing, though it would be good if we could postpone a couple of days. So I contact the travel agent again and tell them that we will take this option but that if anything else turns up with spaces a couple of days later then that would be excellent. A few days pass and we begin preparing ourselves for the rush of having to leave on the last day of term and return on Christmas Eve.

But then suddenly, the day before we are due to leave, the travel agency rings to tell us that they have managed to find places two days later. We can leave for Thailand on Sunday 19 December and return the afternoon on the 26th. We can't believe our luck.

Khao Lak

Thailand, Khao Lak, Happy Lagoon. It is as magical as it sounds. A beach bar on the sand, just a few meters from the bright turquoise sea that stretches all the way to India. A Singha beer in your hand, some planks nailed into pine trees as benches and tables, uncomfortable but perfect, the sun in a deep blue sky, children nagging for a third Coca-Cola, music.

We are sitting at the same bar as Auntie My and I sat at with her children, Gösta and Karin almost three years ago. It is this simple bar that I wanted Bertil and Julius to see too. It is still owned by the same Thai couple. My and I had sat here messaging our parents back home to tell them what a good time we were having. Our mother messaged us back saying how happy she was to hear it and how she would have messaged herself over if she could.

I think of my mother and how she never made it to visit us in Singapore, how she never had the chance to return to Thailand where she had traveled in her youth, and how our time in Singapore did not turn out as we had hoped.

Today it is 23 December and it is exactly one and a half years since my mother fell sick, only three weeks before she died. Ever since then I have tried to come to terms with the fact that my mother is gone and that she never came and never will come to visit us in Singapore. I have tried to accept the fact that our Singapore time cannot be as we had envisaged, with my mother and father coming to visit and then continuing on a trip around Southeast Asia. My mother was so thrilled that we were going to be there and had involved herself in all our plans before we left Sweden.

But then on 23 June 2003, the same day that they were supposed to leave on holiday, my father rang me to say that my mother had been admitted to hospital. They had been on the bus to the airport when she had suddenly got a blinding headache. The bus had stopped and the ambulance had arrived. It was a stroke. On 5 July, the day that we were due to leave for Singapore, she underwent surgery. The operation went according to plan but she died only a week later on 12 July.

After the operation she had been moved from a room with three other patients into a room by herself. I sat beside her, where she lay hooked up to a respirator and fed through lines in her arms. She had a catheter and there were wires linking her to a machine that registered her heartbeat. Her hands and legs were puffy and swollen, her head was bandaged and she had a compress over her eyes. I could see her dark copper-coloured hair under the bandage and could see where it was turning grey at her temples. She lay completely immobile and the only sign of life came from the bleeping of the machine that indicated that her heart was still beating.

I sat on the chair beside her reading a science magazine and waiting for her to die. She waited until I had finished it. I heard how the rhythm of the bleeping slowed to fifty beats a minute, then forty, thirty-five. When it dropped to twenty I ran from the room to find help. I understood what was happening and what it meant and screamed for help even though I knew there was no help to be found. A doctor and a nurse arrived when the bleeping had reached fifteen beats per minute, and then it became just one long, monotone. I grasped my mother's hand in both of mine and cried out loud. She had never regained consciousness after the operation.

The nurse asked me to leave the room for a little while. When I returned, there was a lit candle on the table and my mother was lying peacefully as though she was asleep.

All the tubes and wires had been removed; they weren't necessary any more. I moved close to her, bent forward and kissed her forehead. It was already cold.

For the year and a half since her death I have been living on the other side of the world. We left for Singapore the day after her funeral on 30 August 2003. It had been agonizing leaving my father and My behind with their grief as I traveled to Singapore torn between loss and longing. I peered out of the window of the aircraft into the universe and wondered if my mother was with us. Then I looked inside at my three expectant children and felt easier.

Now, one and a half years later, I feel I have taken a step onwards. Today, 23 December 2004, I feel I can walk with a lighter step even though my mother has left us. It is a relief to be able once again to enjoy life fully and be completely present.

And now I am able to show Bertil and Julius this favourite bar.

Similan Islands

The sun is beating down from a clear blue sky, just as it does every day in Thailand. We are sitting at our table enjoying a long, slow breakfast. There is a buffet with everything you can imagine: English breakfast with scrambled eggs, bacon and sausage, or continental, with croissants and strong coffee and a spread of exotic fruits.

I'm content with a plate of fruit and nuts over which I trickle liquid honey. I could sit here for ever, sucking this atmosphere in. Not far out I can see the water glistening in the sunlight and I can hear the waves softly breaking on the sand. It is only half past eight and the air is still cool. We have the whole day ahead of us. But I want to savour this moment, just as I want to savour every moment in this place.

The mornings are a special time of day, gentle and easy. Below the terrace where we are sitting there are flowers of every colour. There are orchids, ranging from shimmering white to rich purple. The hibiscus flowers that we try to grow at home in small pots grow here into impressive bushes that are covered in blossom. Their petals glow in scarlet and they are shaped as though they are ready to consume the world. There is such a variety of plants. Down by the water's edge it looks like an impressionist painting, generously daubed with bright colours that stand out sharply against the sky blue and grassy green.

High above us in the pine trees we can hear birdsong. It is impossible to see them but we can hear a whole orchestra up there. There must be as many different kinds of birds as there are plants here.

What is it that makes me feel so content? I'm sitting on a bamboo chair with soft red cushions. The table in front of me is spread with gifts from nature: fruits, honey and nuts. I listen to the sounds of the sea and the birds and my eyes feast on all the brightly colored flowers. The temperature is just right but where the sun rays touch my skin I can feel their power.

It is marvellous to sit like this indulging myself and knowing that I have the whole day ahead of me. We have as many days of this holiday ahead of us as behind us now. After breakfast we are going to take a speedboat out to the Similan Islands, which are renowned for their diving and snorkelling.

Although the boat skips along over the water, it takes over an hour for us to reach the islands. When we arrive we are stunned by the sight of turquoise water surrounding dazzlingly white sand, which in turn rims an exotic jungle that climbs up from the beach. We spend the whole day snorkelling and enjoying ourselves.

An incredible day at these islands is soon too close as our boat stops at one last beach before we are to return to our guesthouse on the mainland. Gösta stays in the boat, playing with Karin while Bertil and Julius have taken off with their masks and snorkels. Without a thought, I grab my mask as well, throw myself into the water and swim out. A short way from the boat, I take a deep breath and then dive down among the fish.

It is startlingly beautiful. It is like swimming in a carefully tended aquarium. The underwater plants and sparkling, blunt arms of coral and the multitude of fish – it is as though God has organised a competition in colour and shape. It has all been created with such extraordinary imagination.

I dive down some three meters, feeling the space of the water around me, as though I have entered another world. I allow it to embrace me as I turn to look upwards. I see how the sunlight is fractured into a rainbow by the water.

And then I am out of air and have to return to the surface. As I approach the surface, I wonder how it would feel not to be able to reach the air – the element to which I belong. I break the surface and gasp for breath.

The contrast between the world beneath and above the water is dramatic. As I leave the quiet of the sea that is broken only by the sound of fish nibbling at stones and sand drifting with the movement of the water, the sounds above the water are almost deafening. The waves seem to smash against the rocks, the boat engine roars and I can hear the shrill voices of children along the beach. The sunlight no longer sliced and dampened by the water, strikes me in the eyes with brutal force. I begin swimming urgently towards the boat to check on Gösta and Karin. When I climb onto the boat, Gösta is nowhere to be seen and Karin doesn't know where he has gone. Five minutes was enough for me to lose control over the situation.

I jump back into the water and start to swim out. I turn and thrash through the water – where is Gösta? I feel powerless, diminished. I look about me, scouring the water's surface as though I was lying on the ground, my eyes at ground level. I want to rise up and stand on the water so I can see better. But instead I dive down. The area I must cover seems suddenly to have doubled as I stare through the mask at the eternity of this underwater world. Where is that kid? I want to find him but am terrified of finding him lying at the bottom, maybe ten meters below me.

I curse myself for having loosened my control over him. I continue frantically searching and then I notice Bertil and Julius swimming towards me. I begin swimming in their direction, shouting to ask whether they have seen Gösta. They haven't, so I ask them to help look for him. I turn and swim hard in the opposite direction, calling out his name. I spin in the water, screaming his name in every direction, out towards the sea and in towards the land where massive cliffs rise from the water. I keep diving down to reassure myself that he isn't lying on the bottom and in the confusion of shifting from air to water I seem to forget when to breath and when to hold my breath. There is time for so many thoughts to pass through my mind, so many possibilities: the adrenalin is pumping through me. I must hurry before it's too late. Shit, where is that kid?

And then at last I hear a faint 'ja'. Am I imagining it? Or do I really hear a distant voice? I call out again, this time facing land because it seemed to be coming from there. Now I can hear it coming stronger.

"Jaaaa."

I look towards the cliffs and see a small figure leaping along and waving his arms, crying:

"I'm here!"

Aw Gösta, you are there! The sense of relief at seeing him standing there on an outcrop of rock is absolute. I can feel the tension in my body ease. Thank goodness. I begin swimming towards him and he is scurrying around the rocks now, down towards the beach to meet me. I reach the sand, haul myself up and wade through the shallow water to reach him.

"Gösta!" I exclaim as I hug him to me, "I was so scared when I didn't know where you were!"

December 24

Early in the morning of Christmas Eve we are all sitting together in bed munching expensive chocolate as the children open envelopes from their grandparents. Outside, the sun is already beating down as it has all week to the accompaniment of the constant rattle from the fishing boats. The floor in our room is soon strewn with chocolate wrappings and the children are becoming impatient to start the day.

On our way to breakfast we bump into a Danish couple we've met here and their three-year-old son is charging around chasing a cockerel across the grass. They are staying in the beach bungalow beside ours and Karin has been visiting them regularly to play with their son. We wish each other a happy Christmas and say we'll see each other at the Christmas buffet and the show that is to take place in the evening. All through breakfast Julius and Gösta whisper and giggle together and when we have finished they leap up and say that Father Christmas has arrived.

We rush back to our room but are too late; Father Christmas has already gone. All that's left is an arrow made of small twigs and it is pointing towards the bed. We sneak towards it, lift the cover and find another arrow, pointing towards the next bed so I go over and lift the cover there too. Beneath it is an intricate carving of three wooden elephants. I am astonished. Bertil and I look at one another. However did they manage this? I am so touched – they had planned all this and taken us completely by surprise. I imagine the wooden carving hanging up at home in Sweden as a constant reminder of our Christmas in Khao Lak 2004.

We pack up a tasty Thai lunch box and begin making our way in a longtail boat to a beach that we can have all for ourselves. I chuckle to myself as the boat dips and rises on the water and Bertil tries to capture us with the video camera.

"Imagine next year, when we're sitting watching this film!"

I look into the camera and say to myself, 'Hah, I'll bet you're envious now, aren't you?'

Bertil and I had made a gross miscalculation. Gösta was devastated when he realized he would only be getting one Christmas present. We realized then that at nine years old it's important to get a whole heap of presents. They are supposed to be temptingly piled up under a bushy tree that's full of decorations and live candles. We quietly promised ourselves that next year Gösta would have his dream Christmas.

In the evening I fall asleep with Gösta on one side of me and Karin on the other. Julius is asleep beside Bertil. Before we fall asleep, Gösta and I talk about the day we've had and I wonder if he has enjoyed his Christmas. The mobile phones we've given the boys have a video function and they have been playing with it all over the place, filming this and that. They filmed the magnificent spread at dinner – a mammoth watermelon had been carved into a fantastical monster and Gösta was so impressed that he took a photo with his phone and then set it as his background on the display screen.

I ask Gösta and Karin whether it hadn't been a wonderful Christmas, despite the fact that it wasn't a typical Swedish one and they agreed. I explain to Gösta that it is so much better to get one special present that will give pleasure for a long time to come instead of receiving all those small things that simply break and get forgotten. He only partially assents because really he would have preferred to have more packages to open.

I console myself with the knowledge that the latest Harry Potter film is waiting for him back home.

I never had the chance to give it to him.

The Day the Wave Came

It is our last day in Khao Lak.

The sky is as blue and everything is as quiet and still as it has been on all the other days. There is hardly any breeze. The flight back to Singapore isn't due to depart until eight this evening and we have booked a taxi at four o'clock in the afternoon. That gives us almost a whole day on the beach.

After our last breakfast together on the verandah of the restaurant we walk back to our little cottage along the narrow pathway that is edged with bright flowers. We pack our bags so that we won't have to think about it later in the day. Bertil is feeling feverish so he decides to lie down for a while and join us later. I gather together our beach things and call the children so we can go down to the water. My plan is to lie in the shade and read my book and to have one last massage before we head back up for lunch.

Julius and Gösta are on the cottage verandah, putting on their swimming trunks. Gösta has his yellow Speedo trunks and they are so long that they cover his knees. Julius has blue ones but Karin insists on putting on her

"Royal Dress" – a yellow frock with pink crowns on it. We parade off towards the beach, Julius and Gösta racing ahead of little Karin, who is trotting along, eager to catch up with her brothers, and me lagging behind them. I feel as though I have all the time in the world. Carpe Diem.

There are moments that we know, even as we are living them, that we will never forget. I look ahead to my children and feel so privileged. I watch them as they charge off, all three. They are healthy and happy; they are my children. I swell with pride.

When I arrive at the beachfront I capsize onto a sun chair and prepare for a long, undisturbed read of my book. Julius, Gösta and Karin take-off up a sand dune and begin playing King of the Castle. I catch them out of the corner of my eye and wonder how it will end. Things seem to be getting rough; Julius is dragging Gösta down the sandbank by one leg.

After a while, a Thai masseuse comes up to me and tries to persuade me to have a massage instead of reading. She sits down on a sun chair beside me and we talk. She says that she lives with her family in Phuket and early each morning she hitchhikes to Khao Lak. The money she earns there keeps her, her seven-year-old son and her drunken husband alive. She explains that many Swedish people come here and they are kind and make willing clients. She has an adult daughter who works in business further inland. I am intrigued by her story and ask her to tell me more but she is anxious to work and asks me to come and lie on her massage table.

We get up and as we are heading towards the shade where she has her table Julius and Gösta come running up saying that there is something odd about the water. A powerful current has swept the water back some hundred metres, exposing both the sand and the rocks.

"Goodness," I say, "it's amazing what an effect last night's full moon has had on the sea."

Julius and Gösta ask for some money to go and buy some pineapple, cut like a lollipop and then they run off to find some. Karin has arrived too, and now she tears off to catch up with her brothers. I turn my attention back to the masseuse and lie down on her table.

The massage turns out not be as pleasant as I had hoped. Maybe it's because I've been having massages every day and am tender. I find it difficult to relax properly. But it's alright to just lie there and be taken care of. I think over the past week. It is already over and everything has worked out just as we had hoped: a relaxing, uneventful week filled with lazy days on the beach. After this trip, no one could accuse us of dragging our children off to dangerous places. Now we just have the journey home and then it will be time to start preparing for New Year's Eve. We've arranged to have dinner with some Swedish friends and our Singaporean friends we met recently, Lissy and Joseph. We'll be able to look out over the sea and watch the fireworks being let off on boats along the coast. Like our other holidays, this one has been no problem. Except, I remind myself, we aren't home yet.

All of a sudden my masseuse stops massaging me and begins talking animatedly to her friends who are working beside her. I lift my head to see why she has broken off and notice that everyone is looking at the sea and pointing. I pull myself half up to see what is going on and I see the strangest thing. Far, far out by the horizon it looks as though a cliff has risen out of the sea in a long, broad band. It is brown. I wonder if it was the ebb of the tide that was so strong that it pulled the sea back far enough to expose underwater rocks. But they are so high that one would surely have seen them before. Oh well, I figure, and lie back down.

My masseuse begins kneading my shoulders again but the harmonious feeling I had before does not return. After a short while, she stops again and walks over to her friend, who has begun gathering her things together, and I realize that something is not right. I get up and put on my bikini top, which was lying on the table beside ours.

"Big wave," the masseuse says, pointing to the rocks on the horizon.

"Big wave," I think, but it isn't moving. If it was a wave, it would be moving.

People are starting to become anxious and many are calling to those who are standing on the sand where the water should have been that a big wave is coming and they must come back up.

I gather together Julius, Gösta and Karin and we retreat a few metres up from the beachfront. Some tourists are starting to drag their sun chairs back from where the water's edge used to be. I now accept that it is in fact a huge wave that I can see and I tell Julius to run and fetch Bertil and the video camera; this is going to be something to film!

Julius rushes off and the rest of us begin to wander at a leisurely pace up from the beach, half turned backwards to watch the sea. Many people remain transfixed, gazing at this novel spectacle but others are starting to trot upwards. When we've covered about forty metres, Julius comes running back, clutching the camera and filming. He is giving a commentary as he films, saying there's a massive wave on its way in and running backwards at the same time.

Suddenly, Bertil appears from nowhere. He is standing with legs apart and arms outstretched screaming to us that we must run, quickly. We're now about seventy-five metres up from the beachfront.

Julius turns off the camera, turns and takes off. Gösta is ahead of him and he joins in, running too. Karin is a few metres from me and when she starts to run she trips and falls over. Before I realize that I must help her, Bertil has reached her, picked her up and begun to run with her under one arm.

When I break into a run, I have my whole family ahead of me. We pass by the restaurant some one hundred metres from the sea and it crosses my mind that we should go in there to take cover, but we continue through the whole holiday village and out into the parking area.

Bertil is behind me now, but I can't see Julius or Gösta. I call out to them and then they materialize on my left. I have a stark memory of Gösta in that split second; he is looking at me and there is something surprised and nervous in his eyes. I remember that his fringe was blowing in the breeze. This is the last time I shall ever see Gösta. But I don't know this at the time.

I presume that Julius and Gösta have kept on running and I turn to look for Bertil and Karin instead. They had fallen behind. Bertil was struggling to run with his fever, barefoot and carrying his fourteen kilogram daughter.

In the parking area people are racing around in a panic dodging cars and mopeds. I don't know why we are running or from what. We are perhaps four hundred metres from the sea now and it just seems as though everyone is running for the sake of it. It is as though some threat is being forced upon us, though I don't yet understand what it is.

Another fifty metres up, a pickup truck is reversing from one side, trying to spin around and make a getaway. I turn and call out to Bertil that we can jump up onto the back of the car but I have only time to touch the bar at the back before something hits me from behind and knocks me away from the car. Both the car and I are thrust forwards by a mighty force. Suddenly I am under the car, under water. I am on my back being swept along with the car, my head in the direction we are moving. I shall die. How can I ever get out of here? My body is flung between the four wheels and I realize I am caught in a trap. I don't even try to find a way out; it feels pointless, and I am completely calm. So I just allow myself to float along with the car that I am fastened beneath.

Without warning, I find myself slipping out from under the pickup. I float to the surface and gasp for air. Before I even realize that I have survived, I begin wading away from the car, looking around me. Where are the others?

This is all I care about. Where are Julius, Gösta, Karin and Bertil? I am unable to think of what has happened or of how everything looks around me. I have only gone a short way from the car when I see Julius clutching onto a tree a little way away from me, and not far from him is Bertil who is holding onto both Karin and a tree.

I continue searching the scene. Where is Gösta? My eyes dart from tree to tree and I see people hanging in bunches around their trunks. But Gösta is nowhere to be seen. I spin around and look out over the water.

The water comes up to my waist and is thick with mud. It covers everything around us except for small hillocks where there are trees. I push my way forward between rubble and uprooted vegetation. A snake slips away before me and without bothering about whether it may be poisonous, I continue making my way towards Julius, Bertil and Karin. I can feel fluid seeping from my nose and I lift my hand to my upper lip. Is my nose bleeding? No, it is only water and it keeps on running but I pay no attention to it.

When I reach them, Bertil and I ask each other at the same time, have you seen Gösta? Neither of us have. We tell Julius to take care of Karin and tell them both to hold on tight to the tree. And Bertil and I begin picking our way from tree to tree where people are still holding on. We call out, but the sound of our voices is drowned by the noise of swirling water. But right where we are, the water has halted and not far from us we can see the road beginning to reappear. We run in opposite directions. I am heading towards the sea and the holiday village. Not far ahead, the road is still hidden by the water and an electricity cable is lying straight across it. I step over it, thinking as I do so that there may be a current from the cable but it is as if it makes no difference to me.

Aimlessly, almost apathetically, I begin tracking back and forth along the exposed stretch of road. I look over towards the cottages but all I can see is rubble and overturned cars. I call out to Gösta. My eyes shoot around but I don't know where to search, where to begin, what to do. I still don't know what has happened to us. I know where Julius, Karin and Bertil are, and I know that Gösta is not there.

Only a few minutes have passed since the wave swept over us but my mind suddenly knows with utter clarity that Gösta is gone. I scream for him, but stop in the middle of his name. My whole body crumples and it is over. There is no point in continuing to call his name. He should have been here, but he is not. It is too late. He is gone. Was it at that moment that he died?
My whole being is extinguished.

***

Totally empty, I glance back towards the bungalows and restaurant. There is nothing left of them. I see only the demolished walls and roofs of the reception building and small shops. A car has been tossed up onto a roof. All around on the ground, where the water has now drawn back, are heaps of smashed debris. I can't make out what is there but I realize that houses, cars, machines and televisions have all been churned together.

I turn and run upwards but find only vegetation and a shallow valley that leads down towards the sea. Water has gathered there and is now gushing back to the ocean. This channel is some thirty meters wide and the furious flow within it is like a spring flood in the Swedish mountains; eddies and whirlpools suck at uprooted coconut trees that flash past. The rapids are flowing down from where the wave struck us and I am terrified that Gösta may have been dragged into them. The coconuts that rush past are about the size of a human head and I can't bear to look any longer. Instead I let my eyes be pulled along with the water towards the sea and am appalled by the force with which the waters collide, casting up an ugly white foam. Jagged rocks that were previously hidden by the sand have been exposed and all I can do is stand and stare, riveted to this unfathomable horror.

Moments later my trance is shattered and I start running wildly towards the spot where our bungalow once stood. I am completely disorientated, all points of reference are gone except for the odd sign of a collapsed building. I reach the place where I believe our bungalow must have stood and recognize a tall pine that had stood by the entrance. The tree has survived but the bark has been peeled away several meters up the trunk.

I catch sight of our Danish neighbour and make towards him. He tells me he is looking for his wife and little boy and asks me if I've seen them. He circles round and round but finds no trace.

All that is left of the bungalows is their concrete foundations and some of these have also been torn away. Otherwise, there is nothing.

And everything is silent. I am suddenly aware that we are completely alone. No one is calling for help, there are no injured figures moving about, there is no birdsong. Simply desolation.

Without a word, we begin picking our way over the debris and heading towards the place where the pool had been. Everything has been flattened. We are side by side. My mind observes but is unable to evaluate or take in anything. There is nothing to compare this with.

Then the silence is pierced by a voice calling, calling to us. We turn in the direction from which the voice is coming and see a Thai woman flailing with her arms.

"New wave, new wave, come, come. You have to go away! Hurry!"

I am gripped with fear and I look at the Dane but then begin running. He is soon behind me and on the road further up we are again beside one another, scrambling through the piles of planks and broken metal.

When we reach the spot where the wave hit us we find that most people have already gone. We continue up to the main road and there I find Bertil, Julius and Karin.

"Take the children and go with the others," Bertil commands me, "I'll continue searching and then I'll come up and you can take over."

I peel Karin away from Bertil's shoulder and begin trekking up towards the jungle on the other side of the road. Julius says he wants to stay where he is because he needs to throw up but I scream at him that he will have to vomit while he's walking, we have to get away before the next wave hits.

A few hundred meters up in the jungle I sit down on the earth with Julius and Karin beside me. People are huddled in groups around us. Julius still has the video camera clutched in his hand. He has held onto it through the whole ordeal. People had screamed at him to drop it but he had held onto it anyhow as though it was the only thing that mattered.

It is quiet now save for the whispers of breeze in the treetops and the screech of the occasional bird overhead.

Someone has given me a t-shirt to put on but it sticks uncomfortably to the blood that is still oozing from the wound on my back.

The sun is high above us but the heat is softened by the forest around us.

Bertil is still looking for Gösta and I simply sit and stare blankly ahead of me. My feelings seem to be outside of my body. I am coolly rational, contemplating the fact that we are sitting in a jungle among so many other people. Everyone is so friendly, sharing out whatever they have with them – or rather, the Thais share their goods with the tourists. Karin and I have gathered some banana leaves as a makeshift cover to protect ourselves from the insects and someone has given us a blanket to sit on. Drinks are being distributed and someone has begun bandaging wounds.

The atmosphere is so congenial, and I begin idly thinking about which church Gösta should be buried at. I am hovering in no-man's land. Everything has been irrevocably changed and yet everything around me is so ordinary and familiar.

My thoughts are broken by a knot of people clambering up the slope towards us. I look closer and see Bertil among them, but no Gösta. Bertil stops a few meters below me and shouts that he hasn't been able to find Gösta. He breaks down in tears and cries out that Gösta is gone. I am powerless to move and remain where I am, staring at Bertil in astonishment. What is he crying for? I am so numb that I cannot express anything. I feel nothing and am perturbed by Bertil's exasperation. I am beyond all of this and cannot understand why he is crying – what is going on?

The Chocolate Box

The hours pass. We are sitting upright on our banana leaves and blankets amidst the clusters of people around us. Julius, Karin, Bertil and myself. Most people are silent. No one seems to have the energy to start a conversation and there is little to say anyhow. I know that I too have been back to hunt for Gösta but that I have just wandered aimlessly about among the wreckage.

There are many huddles of people in front of us and some behind us. Then I see an elderly woman weaving her way between the people about twenty meters down the hillside. I can't see what she is doing at first, but she seems to be bending down to ask people something. The closer she comes, the older she looks, as though she belongs to the jungle or another world. I notice then that she is holding something in her hands and that she is proffering it to those she passes. I can't yet make out what it is she is handing out but I can feel myself bristle – there is something inauspicious about this, as though a bad omen is approaching.

When the woman is just a few meters from me I see that she is offering people chocolates and my innards twist and protest. The circle is complete.

She comes straight up to me and reaches forward to show me the exact same chocolates that I received on the day that Gösta was born. I look down into the chocolate box and instead of seeing her chocolates there I see the chocolates that had stood by my bedside in Danderyd Hospital almost a decade earlier. I see a newborn baby in my arms and watch as he tugs hungrily at my breast and feel the chocolate melting delectably against my tongue.

At that time, I had welcomed both the chocolate and life itself, but this time I said no to both.

***

As darkness falls, the screech of the crickets intensifies. Sometimes it is drowned out for a moment by the sound of some other animal – a bird or an iguana perhaps? We are all still sitting here as we have been for several hours now. We have nowhere to go and are cut off from the world. No one has a mobile phone anymore and no one knows that we are all sitting here in the jungle. We can only imagine the dimensions of the devastation. All we know is that the places we had spent the previous night in are now flattened and that we cannot return there. No one comes to find us, so we remain there, letting time pass.

It is no longer possible to discern the groups of people and the only thing that tells us we are not alone is a gentle murmur. A Thai man has lit a small fire beside us and two shadows arrive to join him, two more Thais. They are carrying a newly caught squid that they thread onto sticks and then hold over the flames. They invite us to taste some, but I decline. I gratefully accept the water they offer, however. It will be four whole days before I can eat anything again.

"It's as though I sensed that something was going to happen," says Bertil out of the blue. "When you were all down on the beach I was lying in bed playing patience. I completed four games but not the fifth."

Just above us, two Thai women have arrived and settled themselves down. They lean forward to look at my blood-stained shirt and mutter in horror. They speak reasonable English and they tell us that many people left the area in various vehicles during the afternoon. We explain that we had been staying in the bungalows at the foot of the hill and that we were supposed to be on a flight to Singapore right now. I am struck by the realization that we have missed our flight home. I wonder whether the aircraft is standing waiting for us or whether it will just take off with five empty seats.

The Thai women continue asking us questions and we tell them that we have lost our son. They become agitated.

"Your son? How old is he?"

"He is nine."

"How does he look?"

"He is blond with blue eyes."

"You mean blond like him," they say pointing to Julius. "Does he look like him?"

"Yes, kind of."

"But we saw a boy who was blond and skinny and about nine years old down by the road. He had hurt his foot and could not walk. Someone carried him and put him on a pick up."

The women carry on convincing us as we compare their description with Gösta's appearance. Well, maybe. Maybe it could be Gösta. A flicker of hope sparks to life against all odds. Logic begins to battle against conviction; of course it is possible that Gösta has survived and ended up somewhere else. But the deep-seated feeling will not yield.

Bertil and I turn to each other: hope or despair? We start discussing whether or not it could be Gösta and I become eager to take off and look for him. But the darkness is compact and there is nowhere to search just now. I have to wait until morning. Should Bertil go off alone? But we must drop that idea too. He has neither shoes nor glasses now and we don't know how we would be able to contact each other the next day. We can do nothing. So we try to comfort ourselves with the notion that Gösta has survived and is being cared for. But I see him before my eyes, alone and injured and I feel so badly for him. He must be wondering where we are and why we aren't coming to find him. He must feel totally abandoned. Please let this night pass quickly.

It is silent now. Everyone seems to have drifted off into their own thoughts. I clutch onto the image of Gösta together with strangers. It is terrifying.

After some time, a long-haired young man approaches us. He is Australian and I realize it was he who first screamed the word tsunami. I've never heard this word before and I don't bother to take note of it. I have no idea that it is a word that will become etched into my memory for the rest of my life.

The Australian explains that there has been a massive earthquake that has affected large parts of the world. I start asking him questions but he says he is on his way to another collection point to inform others and he hurries off.

Bertil and I look at one another and think the same thought.

"Wait!" we both say at the same time. "You have to look for our son. He has been picked up by someone and brought to another Rescue Center."

"Yes, of course I'll look for him. What's his name and how old is he?"

"Here, I'll write it down for you. His name is Gösta and he is nine years old."

Bertil manages to get himself a piece of paper and a pen and writes a note to Gösta: 'Gösta! Mamma, Pappa, Julius and Karin are ok. We'll come and find you tomorrow. We love you!! Mamma and Pappa'.

Bertil hands over the note and says Gösta's name out loud, describing what he looks like. The Australian takes the paper and puts it carefully into his pocket.

"I promise to send him your love and tell him you're okay and will find him tomorrow. Bye."

He turns and is gone into the darkness.

Silence again. Bertil lies down with his arms behind his head and seems deep in thought. I remain sitting, my eyes pinned onto the place at which the Australian had been standing. I see Gösta again. How badly injured is he? Is he in pain? Is he afraid? Distressed? Yes, of course he is! He's missing us. He wonders where we are and is anxious. Is anyone with him who can comfort him?

The knowledge that he may be somewhere not far from us intensifies and I begin to long for him much more intensely.

In front of me, Julius and Karin have managed to fall asleep. Bertil has turned onto his side so that he has his back towards me, but I remain sitting upright. I've given up the idea of trying to lie down. I can't lie on my stomach or back because of my wound and anyhow I don't want to lie down.

In my mind, I can still see the Australian's back disappearing into the jungle and in my imagination, I follow him. I can see him arriving at the next place, asking for Gösta, finding him.

I hold onto this thought as I look out onto darkness – darkness that stretches upwards and away. I see a tree trunk and my eyes follow it up to the crown that is now only just visible against the night sky, and then I see my mother there. 'Damnation, shit,' I burst out. 'Don't let it be true that you are the one who's greeted Gösta and are comforting him.'

December 27

After an interminable night in the jungle, broken only by the light of a few fires and the full moon, everyone who has gathered there begins as day breaks to make their way down to the road again.

The previous evening I had discovered two wounds on my foot. They were deep and full of dirt and now it had become difficult to walk.

Descending from the hill is difficult. Although there are pathways, they are edged with sharp stones and in places it is so steep that I have to put one foot on either side of the path to balance. But after about an hour's descent we reach the road where we find a pickup truck with the back open. There are already people sitting in it and we crawl up as well. When it is full, it begins to move.

It turned out it was not so easy to hunt for Gösta. Like fools, we had imagined there would be only one collection point, but there are many and it is chaotic. A few tourists and Thais have begun writing lists of the names of those missing but no one knows what anyone else is doing.

We change trucks a few times and drive past mile upon mile of desolation along the roadside. Here and there the road is blocked by heaps of wreckage and in places we see lifeless bodies and upturned vehicles. It is still impossible to take in the enormity of what has happened.

We eventually arrive at a temple and our Thai driver makes a tight turn to the left before drawing to a halt on the gravel. We are a group of tourists sitting in the back of a pickup truck.

Suddenly someone is careering towards the truck, bellowing at the driver. The driver flings open his door and bolts, calling back to us in a language we cannot understand. We look at one another and then around us. Now we see others running upwards from the temple area and without thinking, I grab Karin, leap down from the truck and scream at Bertil and Julius to follow. We drop the few things we had with us, including the film camera that Julius has managed to hang onto until this moment. We join the hoard of people going higher, higher. I am stricken with the thought that we are abandoning Gösta, running without him.

We find ourselves in some kind of animal park. There are monkeys in cages and other animals. I feel so sorry for them but we continue, over the barbed wire fence. Julius gets caught on it and it rips his skin, but we keep going. We reach a road that leads on into the jungle and then Bertil stops in his tracks.

"Wait," he says, breathlessly. "Take it easy. I don't think there's any danger. We can't keep running like this."

"No, come on! Everyone is running!"

"I assure you. It's not possible that another wave is coming."

I stop then and Julius and Karin look questioningly at us. We have stopped just outside a small house where a Thai family stands gathered. The man of the household comes towards us and invites us into his house. We gratefully accept and sit ourselves on the chairs he points towards. He offers us drinks. We are out of breath and exhausted. I look warily out towards the road and am unable to let go of the thought that we have left Gösta behind.

The family tries to talk to us and we struggle to understand. The woman notices my blood-stained shirt and goes to look for her first aid equipment. When she returns, she tries to help me off with the shirt but it has stuck fast to the wound and it won't come off. I gently but firmly refuse her offer to continue trying and a few minutes later we thank them for their hospitality and explain that we have to go.

People are returning now to the temple area and we join the trail. The truck we had been driven in is still there and in it is the film camera. We collect it and then make our way to the temple. We stay there all day, watching as truckloads of people are delivered, one after the other. Other vehicles arrive at another part of the temple but those contain the bodies of the dead; they are laid out in rows on the floor of a large hall. Bertil forced his way into this hall to look at the bodies but he didn't find Gösta there. I don't want to search through the corpses. If I should chance upon Gösta among them, it would be too late anyhow.

In one area of the temple, Thai medical personnel are caring for the wounded. Later in the day, I limp my way between the injured - men, women and children, on gurneys, on chairs - to have my wounds attended to.

I see a teenage girl lying on a stretcher with one leg raised in the air. The skin has been torn away and the edges of the wound look swollen and angry. But when I look again I see that it is not swelling; fistfuls of gravel and small stones are stuck under the skin and a nurse stands bent over the girl, painstakingly picking out stone after stone. I look at the girl's face but she is staring out into nothingness. The leg and the face don't seem to belong to the same being.

When my shirt has been removed and wound cleaned out and bandaged, I make my way up to the large hall on the second floor of the temple. All along the walls are people sitting or lying listlessly. It is quiet as I walk over to where we have settled. Julius and Karin are sitting on a mat playing cards with some other children.

Suddenly I feel a presence behind me. Someone is trying to get past me on my left side. I step to the right to let them through and I turn to look at them, and smile. I wait for a fraction of a second and wonder why they are hesitating to appear. But no one appears. I stop and turn further, but there is no one. I am alone on the pathway across the hall. I turn around completely and look about me but there is no one there.

I know this has happened before. It was when I was walking along the beach in Malaysia. I had walked off by myself to get a massage and was on my way back. I had suddenly had this feeling that there was someone behind me and all of a sudden Gösta had jumped out like a jack-in-the-box. He had secretly followed me off to get the massage and been sneaking behind me all the way back.

Is he sneaking up on me again?

Later in the evening a car leaves for Phuket. It is rumoured that they are trying to gather all the tourists down there so as to organize emergency help. We reason that it would be a good idea to take this transport to town. The vehicles have stopped delivering survivors and corpses to the temple and we are beginning to feel restless. Gösta is not there so we must move on.

Earlier in the evening we had met someone we knew. He and his son had survived and they were looking for the rest of the family, a little brother and the mother. And we bumped into the Dane who still had no news of his wife and little boy.

It is long after midnight by the time we reach Phuket. We are dropped off at a large gathering point in the center of town. Even though it is so late, the place is a hive of activity. Each country has set up a temporary stand where you can register and report the names of those missing.

We bustle through the people until we find a small stand where it says Sweden. A single young man is sitting there, writing lists. He says he heard on the radio that volunteers were needed to help trace those who were missing and since he had not been affected, he decided to sign up.

We write our names on a list. A Thai volunteer sees my name and calls out when she sees it:

"I recognize that name."

"Hm?"

"I am sure I saw that name at the office where you can call in to report people. Someone has called for you!"

"But who could that be?" I wonder, looking at Bertil and the volunteer.

"I don't know. I did not take the phone call. But if you come with me we can go over to the office and ask."

Julius and Karin have stretched themselves out on the ground and fallen asleep by now. We decide that I will go with the volunteer and see who was looking for me. We are sure it must be about Gösta.

When we arrive at the office, the staff there are just about to leave and go home. My volunteer starts looking for the list of incoming phone calls but she can't find it. She asks the remaining staff if they've seen it and they refer her to someone who was working there earlier this evening. When she explains this to me I ask her to find out where the person is now and whether we can get hold of them. I am becoming increasingly frantic as I realize that many of those who have phoned this office are victims trying to trace one another.

No one seems to know where my name has been registered so I beg them to contact the person who received the phone call. Finally, the volunteer manages to get hold of them and find out where the list is. As this is happening around me, I imagine various scenarios in which some kind person has taken the name of Gösta's mother and phoned around to try to locate her.

Gösta, we are on our way now! We'll soon be with you!

I am allowed to read through the list of incoming phone calls, but I can't find my name anywhere. The volunteer admits that she must have mixed it up with someone else's.

That night, we sleep in the gymnastics hall of a school together with some forty other Swedes. Gösta is not among them either and we are forced to leave yet another place behind us without him.

December 28

Before leaving the next morning we sit waiting for everyone to gather. All of the Swedes who have slept in the gymnasium are going to be taken by bus to the Swedish consulate. Julius, Karin, Bertil and I are sitting together on a white, swirling iron sofa and I have this strange feeling of normality. Everything looks normal around me. Bertil and I are chatting, Julius looks tired and Karin is bored, waiting for something to happen or to see Gösta emerging from round the corner. Everything seems too ordinary. Or IS it ordinary?

I have left my body and am hovering overhead, looking down on this ordinary family from above. It looks pleasant enough. They are sitting there, relaxed, on a white sofa and seem to be unhurriedly waiting for something. They are probably enjoying the sunshine and warmth. Perhaps they are planning how to spend the last days of their holiday before returning to school, work and everyday life.

What a disgusting trick I am playing on myself! How can we be sitting there so calmly as though nothing had happened when we have no idea where Gösta is? How can we remain so still?

I am back in my own body and feel myself standing up to see if the bus has arrived so that we can get going.

Two more days were to pass before I could weep or eat.

Downtown Phuket

Bertil lost his glasses in the flood and has to get a new pair. Up at the provisional Swedish consulate, Anki has offered to let Julius and Karin use her room. Anki has arrived from Singapore, where she works at the Swedish Embassy. We barely know her. I had met her in August at the annual game of rounders that is organized for the Swedish community. I remember her sprinting round the field to reach base before the ball but how neither she nor anyone else could take it too seriously in that heat. Gösta had been with us and we must have been so carefree. It is absurd now to see Anki here and at the same time to hold this image of the rounders match in my mind's eye. Now she is working at double speed to try to help all these victims.

There is a steady stream of people arriving at the consulate. They are desperate in their borrowed clothing, with their wounds and the horror in their eyes. Everyone is empty-handed and questioning; where should one begin, what is the next step, where are our relatives? Anki helps to check the lists of survivors. She organizes temporary passports and tries to assist everyone with the next thing they have to do.

After leaving the children in Anki's room, Bertil and I leave for the hospital to get our wounds dressed and continue our vain search for Gösta.

Bertil has a painful foot and is waiting for an x-ray, so I leave him to go and get my wounds dressed.

On the way back to Bertil, I pass room after room filled with gurneys and mattresses wherever there is space. My rational self looks for Gösta and my inner conviction is quieted for a while. I scour the faces but find no one I recognize. And then I look into one room and see a small person sitting with his back towards the door. His back is the right size and he has blond hair with streaks of strawberry gold and it is longish with a gentle wave. I stop and stare. No. Something tells me it isn't Gösta but for a split second I cling to the hope that it could be him. It could be ...

I enter the room and walk towards the child and don't register that the child a girl who is speaking German, probably to her mother. I have to walk right up to this child and look straight at her face before I am prepared to admit defeat. The mother and the girl both look up at me and I nod cautiously before I retreat to the door, to the truth.

As I continue back towards Bertil I hold onto the image of the small back and the golden hair, trying to feel as I would have if it had been Gösta. Imagine if it had been him sitting there. It might have been. He could have been found and taken to hospital. He could have.

Then it becomes too painful to think of and I banish the thought from my mind.

We leave the hospital and return to the center of Phuket where we find an optician's shop. While Bertil is trying out glasses, I wander through the shop and out onto the street. It is a narrow street, filled with life and small shops. Cars careen along the middle as cyclists dodge about. The sidewalks are packed with tourists and locals sauntering along as though nothing has happened. Perhaps it hasn't, for them.

I don't know what it is that prompts my powerful reaction but it suddenly hits me that I am in this terrible situation and I begin howling like a child. Here I am waiting for Bertil while one of the worst things that can happen to a human being has just happened to me; my son is gone, probably dead, and I'm standing here in the midst of all these people as though nothing has happened!

I could just as well be a mother waiting for her three children to nose around in the souvenir shop next door. I might just as well be for all the people around me; in their eyes I must just be one more person standing waiting on the sidewalk outside a shop in a lively street.

The atmosphere at the hotel we are sent to contrasts sharply with the busy confusion at the consulate. We are to spend the night at a hotel that far outdoes any we've ever stayed at before. It is the Marriott Hotel, just a couple of kilometers from the Center.

Late that evening, we had said goodbye to Anki who kept on working without a break. We got into a minibus that was to take us to our destination. On board were other Swedes and I began to make conversation with them. Within minutes I find myself in a hot dispute with a young man about belief in the Devil and Nazism, which he apparently admires. I don't understand why I have got involved in this outrageous discussion – it just happens.

The room we are shown to is elegantly furnished in heavy Asian wood and the bathroom is luxurious and beautiful. Julius and Karin throw themselves onto the beds, but their exhaustion at once lifts. They burrow beneath the soft down covers and within seconds they have got the TV on and are watching *Shrek*. They straighten themselves out in the bed beside Bertil to watch a film that the whole family had watched together just a couple of weeks earlier.

I slip into the bathroom to rinse off the worst of the day's filth. I'm longing to take a proper shower but don't dare with all my wounds, so I take a flannel and wash myself down best I can. I catch sight of the birds nest that my hair has become and know that I have plenty of time to gradually tease it apart. There is nothing to hurry for.

It is only when I turn off the water that I hear the music. My stomach tightens like a fist. It's the easy-going music from *Shrek* and I think of how we had all been listening to it only two weeks before. The music leads me back to that day, sitting in the cinema with Gösta so alive beside me, downing the popcorn that he is sharing with his brother and sister. My throat tightens and I want to run away. I can't bear to have to sit here listening to the music outside the door. I can hear the others laughing at the donkey's comments to Shrek and I feel the ire rising within me. How can they sit there laughing at the very film that we had watched with Gösta? Don't they understand? Don't they realize that they can't sit there laughing at this film when Gösta has gone? He is gone – can't they see that?

I turn on the tap as hard as I can and shut the door that has been ajar. I don't want to hear any more of this. I feel crushed, caught in a corner that I can't escape. After a while Julius comes into the bathroom and asks why I didn't want to watch the film with them but before I respond he catches sight of the deep bathtub and asks if he can have a bath. I am relieved to have escaped the need to answer his question and tell him instead that it's an excellent idea for him to have a bath and scrub off all the dirt.

Julius sets to treating himself. He finds a little cloth bag of bath salts in a porcelain bowl on the edge of the bath and he carefully pours them into the water. He lights all the candles that are placed on wooden trays around the bath and when the bath is almost full, he lowers himself gingerly into the frothy water. I hear how he sighs luxuriously.

'He deserves this,' I reflect. But then Julius turns to me and exclaims:

"Gösta ought to be here now. I feel guilty lying here in this water enjoying myself when we don't know where he is."

***

It is dark and quiet in the room. We're all in bed and the children are already asleep. The crickets are playing a lonely, melancholy symphony outside the open window. Beyond their cries is only silence.

I find it difficult to get into a comfortable position because of my wounds. I have propped cushions against my back and under one leg but I can't sleep tonight either. I don't expect to be able to. Why should I be able to sleep? Why shouldn't I be able to sleep? Why should I sleep, why shouldn't I sleep? It makes no difference. I just let the time pass and follow it as I'm supposed to. The darkness and quiet outside are signals that we should go to bed, so we have.

But then Bertil gets up from the bed and walks over to the window, a bow window that goes all the way down to the floor. There are two brightly colored pyramidal Thai cushions on the floor on either side of a low hardwood table. Bertil sits down on one side and I can trace the contours of his face in the dull light. I lie still and feel the air, motionless around me. I watch as Bertil stretches out one hand over the table and I listen to the crackling sound as he strikes a match. The room lights up and I can Bertil's face becomes visible for a moment. I see in his features something new, something I've never seen there before: a void, as though something is missing.

He moves the match towards the candle on the table, lights it and then blows out the match. The smell of sulphur wafts towards me and conjures memories of Christmas and St. Lucia, but also of churches with icy stone floors and proximity to an eternity I have never known.

Bertil leans back into the cushion and turns his face away from me towards the window. I don't move, but simply lie there and watch. I feel removed from this and I don't want to disturb it. The candle flame dances softly beside Bertil as he gazes out at the night. I cannot see his eyes but I know that they are fixed on the darkness; it feels as though it is I who am sitting by the window.

I can see what Bertil is feeling. Little by little the fact that Gösta is no more is sinking in but Bertil is calling out to him in his heart, he wants to reach him, he has lit a candle so that Gösta will see that his father is sitting waiting for him. So he'll see that we're all here in this hotel room waiting for him to arrive. It is late and the children have already gone to bed and Gösta should be here in bed too by now.

Bertil is trying to comprehend the fact that he is sitting waiting for Gösta and that Gösta will not return. I don't want to break in. I lie watching as the darkness unfolds something interminable, something that vanishes into a distance we cannot reach. And I see a tiny flame burning in a futile effort to vanquish the darkness that has taken Gösta. The light is incapable of bringing him back; I see it gradually dawn on Bertil that Gösta is out there and that he will not come back.

***

Julius has just dragged us back to the hotel room. Karin has found a playroom and some new friends. She is exhilarated with her new-found freedom to roam the hotel alone. Bertil has gone off to continue the search for Gösta. We have been given information about a nearby hospital where there are many survivors and a temple where corpses have been delivered. When Bertil heard this, he was unable to sit still so we found him a taxi that could drive him around on his quest to find Gösta.

A nurse has just left the room after re-bandaging my wounds. They are still ugly and I realize now that the one on my back ought to have been stitched but that it's too late now. There is something pleasant about having the sores dressed; it stings and pulls but it is a pain over which I have control. It is straight, uncomplicated, not edged with anxiety. Dealing with this pain focuses my attention on something else, if just for a moment, and it helps me forget the consuming fear of knowing that Gösta is gone.

I can't say that he is dead because I don't know that. Reason and emotion continue to battle with one another. I feel certain that he is dead even though I know that he could be alive and could be close by.

It feels good to have a painful wound on my back.

Julius bounds up onto the bed and puts on the TV. He flicks through the Thai channels until he finds an English one with cartoons. It is morning and the room is still cool. I walk over to the balcony door and push it open. A mild breeze brushes past me into the room carrying the scent of freshly watered grass from the gardens below. And I can smell the sea and dry sand. When I look down between the trees and shrubs I see the blue of a pool sparkling in the morning light.

The telephone rings. I turn and Julius' and my eyes fix on each other in surprise. I pick it up and hear a man's voice asking in broken English if I am Ann and I say that I am. He tells me that Gösta has been found.

I lower myself onto the edge of the bed and ask him slowly to repeat what he has said. Again he says that Gösta has been found. I ask where and he says at the Rescue Center. I ask how he knows this and where he is calling from and it is then that I understand that I am speaking to a telephonist in the hotel. I tell him I'm coming down immediately.

I am frozen, phone in hand, eyes pinned to the wall. I am unable to take in what I've just heard. As I replace the receiver I turn to Julius who has crept up beside me on the bed. He looks questioningly at me and I explain to him that they say they have found Gösta and that I must go down to the reception to find out more. Julius says nothing and just looks away.

I leave him on the bed and go out into the corridor. I wince as the sores on my foot make me hobble, but I try to speed up my step best I can.

The image of Gösta sitting on Bertil's knee is palpable; he is seated on his father's lap with both legs to Bertil's right, his arm around Bertil's neck as Bertil wraps him in his arms. I wonder why Bertil hasn't called to say that they have found Gösta but this image tells me. He can't call because he is crying too hard. He has buried his face in Gösta's neck and can say nothing and a sympathetic volunteer has managed to locate me to deliver the news.

It's alright, reason triumphs – of course Gösta has also survived! But these instants of triumph are always brief. Persuasion demands proof. I had omitted to ask on the phone whether Gösta had been found dead or alive. I speed up my step, barely breathing until I reach the reception desk. My head and heart are at now at complete loggerheads.

I ask to speak to the telephonist but am told he isn't available. So I explain what I have been told and ask the receptionists at the desk if someone could contact the Rescue Center. A commotion ensues since no one knows which center to phone and after some dithering and squabbling, one of them goes to fetch the telephonist. When he emerges I gather myself in readiness to pose my question.

"Was he dead or was he alive?"

"Dead."

"No, I mean was he dead or alive?"

"No, he was dead."

"Was he dead?"

"Yes, he was dead."

I keep repeating the question a few more times, unable to fathom what he is saying to me. Finally, I turn away from him and pin my gaze on the ceiling, on the wall, on a painting. I follow the lines of the painting with my eyes, observing how they cross and interlock. The lines describe a road edged with houses and there are two men by the roadside. They are talking to one another. The painting seems naked somehow. It is colourless. The paper is white and the lines, black.

***

I don't remember how I got myself back from the reception area. I don't know how I got through the rest of the day.

Later that day I know that my father and sister arrived. My sister, Auntie My, had managed to get tickets on a half-empty Hercules aircraft bound for Phuket. We are sitting at one of the bars in the hotel and I hear myself telling them that Gösta has been found, dead. They look at me and tears begin to roll down their faces. I see their tears and wonder again how it is possible to cry.

Karin runs up to us and sits down beside me. She is just about to start telling us something but then she halts, looks at each one of us in turn and then her small body slumps. She knows. I take her hand and turn to her, look at her.

"Yes, I say. Gösta is dead."

Karin looks back at me, defiant. She says nothing. I keep tight hold of her hand and look into her eyes and find anger in them. We sit like this for a time, as if in a duel. I cannot surrender. I cannot take back what I have said. Gösta is dead, Karin. She continues staring into my eyes until she is no longer able. Then she suddenly withdraws her hand from mine, looks away from me and springs up from the sofa. I watch her as she disappears behind a pillar and then is gone. Still looking beyond the pillar, I tell the others that I have been in touch with Bertil and explained to him that they believe they have found Gösta. Bertil happened to be close to that Center when I spoke to him so he had hurried over there to find Gösta's body, but there was none. We were back to square one. Had they found Gösta? In that case, where was he now? Was he really dead?

I haven't the strength to continue engaging in this. I let my thoughts and feelings wrestle again and try to keep myself apart from them. I slouch into the armchair and do nothing until Julius comes and plants himself beside me. Do I tell Julius that they think they have found Gösta dead? I don't know.

Then I remember the bag of sweets that Auntie My gave to Julius and Karin. Oh well, they can split the sweets between the two of them now instead of saving some for Gösta, as Julius had insisted.

I turn to him and ask what he would like for lunch. Everyone joins in the discussion about where to eat and we decide on an Italian restaurant. I am about to eat my first meal in four days and when I tell Julius, he is thankful.

***

Bertil has given up searching through the bodies at the temple to find Gösta. The bodies are now so grotesque that it is impossible to recognize who they are. We are sitting together in the dining room eating breakfast. Julius and Karin are devouring pancakes with chocolate sauce with utter disregard for our principle that one should eat something nutritious, like bread and cheese with some vegetables first. What does it matter?

Auntie My has helped the children with their plates and has put a cup of tea and some toast in front of me. I can eat nothing else from the buffet. My father is sitting opposite me nursing a cup of coffee. He has no appetite either and is making heavy work of a sandwich. My finally sits down and Bertil begins to tell us how he searched the day before.

After combing through several hospitals he had gone back to the temple. He had been supplied with high rubber boots and a small plastic tub containing strongly smelling eucalyptus oil. Then he was shown into the spacious halls where the bodies had been laid out in rows. The heat was so intense that the air was quivering and the odour was dense. A volunteer accompanied him up and down between the rows and Bertil tells us how the two of them had waded through worms on the floor. Since it was pointless trying to identify Gösta by looking at faces, Bertil decided to look instead for his yellow Speedo trunks. The ones he put on in the morning outside our bungalow in preparation for a day of swimming and playing on the beach – an eternity ago.

By now, it was impossible to distinguish man from woman, Westerner from Asian. The bodies were already distended and bluish black. Tongues were protruding from the mouths. Some lay with their eyes open, others with them closed. Some were glaring through glasses into oblivion.

Bertil hadn't found any yellow Speedo trunks. But while he was there, there was a steady stream of trucks reversing up to the temple doors so that volunteers could lift down the bodies from the back. The bodies were wrapped in green plastic covers that were bound with rope. They were carried into the temple and then numbered. The previous day transporters had begun using refrigerated containers to transport the dead to try to slow their decay. Bertil could only bear so much and when his search proved fruitless he decided to get away from the temple.

We are sitting at the table and no one is speaking now. Karin has got down and is playing now with a new friend, a little Swedish girl who has lost her twin sister. Julius is still with us, listening. I don't know how to handle this situation, his presence at this conversation. This is nothing for an eleven-year-old child to be listening to. But it is our reality and Julius has a right to know, however appalling it may be.

***

We are back at the consulate to collect our new passports. It is just as disorganized as it was two days earlier but there are journalists and a TV team here now. They are trying to speak to as many as possible but I manage to sidle my way past their microphones.

The Swedish foreign minister blusters past me a few times and the police and rescue services are trying to install themselves. I take the chance to speak to the Swedish doctor there and he administers agonizing cleansing tincture and then advises me to get myself home as soon as possible. The wounds are not looking good and with ineffective antibiotics and a climate that is ideal for bacteria to thrive, it is risky to stay here. I become uneasy and tell My and my father that we must arrange to get tickets back to Singapore as soon as possible.

Singapore, January 2005

We are able to get seats on a flight back to Singapore the next day. We leave Thailand on New Year's Eve 2004 and land in Singapore without Gösta, knowing that we may never see him again.

Our seats are in the fore of the aircraft on the left side. My sits beside me but I don't remember talking with her. We are collected. At one point, I desperately need to pee and I stumble along the aisle towards the toilets in the first class section – as an economy class passenger I shouldn't really be using them. But my foot is so painful that I decide to use the toilets that are closest.

When I come out, an air hostess snaps at me that I am only allowed to use the toilets further back. I nod and hobble back to my seat. Later on, she comes past and informs me that since I am injured, I am allowed to use the first class toilets, and then she does a double take and apologises when she recognizes me as the person she admonished earlier.

Then the captain's voice crackles through the speakers that we are preparing for landing and passengers must fasten their seat belts. I sit back and glance out of the window into the sky. The plane has rolled to the right and I see only blue but as it straightens up I can see all the cargo ships queuing up along the Singaporean coastline.

I am not at all prepared for our return. Through the airplane window the six tower blocks where we live seem to appear in front of me without warning, as though nothing had happened. They are as tall and pompous as ever, waiting for us to come home. Unfazed, they welcome us back after our trip to Thailand. I see our building and can almost see our apartment windows, twenty-seven floors up, looking out of the trees and parkland that leads down towards the beach.

I am not prepared for the chasm that opens before me. Seeing these buildings makes me truly realize for the first time what has happened. Until now, we were in a foreign place, removed from our own environment, but now we are back in the world that has become ours. Gösta had left this life two weeks earlier and he is never to return to it. Now, we are returning to this life having left him behind. And the buildings are beckoning to us.

As we exit the plane I see that a wheelchair is waiting for me on the tarmac. It is a relief to sink down into it. Karin hops up and sits in my lap and Julius pushes us through the terminal and out through customs. We pass the luggage belts without stopping – we have nothing to wait for.

As we come out of the customs area we are greeted by a cluster of people with TV cameras and sound equipment. It takes a second or two before I register that it is us they are waiting for. They move forward hesitantly and ask if they may pose a few questions. Bertil and I look at one another and wonder how they know, and then nod affirmatively towards the reporters. Since I am clearly wounded, they begin by asking if I can explain what happened.

I begin by presenting a factual account but when I come to the point where I am telling them that we have had to return to Singapore without Gösta and that we don't know where he is, I break down. Julius is still standing behind me and I can feel how uncomfortable he is becoming, shifting from one foot to the other and fiddling with the handles of the wheelchair. Karin turns her face to mine, clasps my neck with her thin arms and burrows her face into my chest to hide. And I am sitting their grinning at these reporters as I tell them that Gösta is gone. They turn off their camera.

Bertil's colleague Per and his wife Agneiska have come to meet us. They have kept their distance but when the reporters have thanked us and begun to move away, Per and Agneiska step forward. We follow them over to their car and they drive us home. It was they who met us when we first arrived in Singapore from Sweden, and once again they have prepared for us by stocking up with some milk and bread and cheese. I don't remember anything else about our return to Bayshore.

***

We have been back in Singapore for three days now and are sitting at a restaurant not far from Julius' and Gösta's school. Julius has started back at school and Bertil and I have just collected him and we are having a late lunch. After lunch, we are to go to the Swedish Embassy to organize new passports. A waiter has just cleared the table and is brushing the crumbs of the tablecloth and Karin is licking frenetically at her ice-cream which is melting faster than she can eat it.

My father and sister are relaxing in their large wicker armchairs, each with a coffee and Bertil and Julius have just left us to go and get their passport photographs taken. There are some plastic bags on the table – Auntie My has been off shopping with Karin for new clothes and shoes to replace those she lost in the tsunami. Karin has been upset about losing her brand-new yellow dress with the pink roses on that she had been given just before Christmas and Auntie My had promised to find her another one, just as pretty. And now we're sitting here after their shopping expedition.

We have sat on the large terrace of this restaurant so many times before eating a light lunch with beer or soda. It is popular with tourists but we have always liked it. It always feels cool here and one can sit and watch the heat-weary people on the sidewalk below hurrying by with their shopping bags.

It feels unreal to be sitting here at this same restaurant, where everything is just as before. Behind us is a rowdy crowd of Westerners with carrier bags from some of Singapore's more exclusive shops on the floor by their feet. The waiters glide past in their white uniforms balancing round trays of drinks and steaming plates on their hands. This is an untroubled, leisurely oasis in the midst of the city.

When we were still in Thailand after the tsunami it was as though we existed in a bubble; everything was unaccustomed, surreal. All I had to do was remain focused on the present moment, unintruded upon by memories from my surroundings. But now we are back into the everyday life that we have been living for the past year and a half. Singapore has already become charged with familiarity.

I have had enough and want to get away from this restaurant. I've already had my passport photo taken and on account of my foot, I was permitted to skip the embassy visit and take a taxi home instead. I spring up from the table and declare with urgency that I have to go. My father gets up as well and comes to help me. With linked arms, we make our way out onto the street and flag down a taxi. Before I know it, I am inside, directing the driver.

I have to get home and be alone. I can't bear to collapse in front of the others. I can't deal with their anguish at seeing me like this. Sometimes I just need to cry by myself. As the taxi moves off I let go and without making a sound I allow the tears to flow in an unbroken stream for the entire journey back to the Bayshore apartments.

Bayshore. It is incredible how much a name can mean, how many different feelings it can contain. This name used to smack of pride and coolness. It was outside of the normal – a different kind of address from the one I had grown up with. I had always known that our stay in Singapore would be an extraordinary, bracketed pocket of life, something different from our normal lives. I had been sure that it would eventually become a fond memory. But now I realized that Bayshore was to become a memory with a very bitter aftertaste and mixed feelings.

In the elevator, on the way up to our apartment on the 27th floor, I realize that I don't have my keys with me, so at out floor I get off and find myself sitting outside the front door, waiting. A threshold of a front door. It can mean so much. This was where we had once stood, the same evening that we had arrived from Sweden, waiting to see our home in Singapore. We were curious to see the place that Bertil had found for us, and inside was a new bicycle for Gösta and a stereo for Julius. The front door was made of dark wood and it had a brass door handle. Inside, it had marble floors that were speckled in brown, beige and terracotta. It felt classy. Immediately to the right of the front door was a high steel grid and railing that you could look over to see straight down, a giddying twenty-seven floors. At the foot of the door was a small ledge on which we would later place our shoes on a wooden IKEA shoe rack.

Now I'm sitting here looking at Gösta's shoes on the rack, which are waiting for him. It's a pair of red and white running shoes and the laces have been knotted so that he can slip the shoes on and off without untying them. They won't be needed any more. I am wailing now, able to let the grief free, uninhibited. I see the elevator from which two boys in school uniform have emerged each day before kicking off their shoes, putting them on the rack and ringing the doorbell. On the other side of the door was a mother in an apartment with panorama windows offering a spectacular view of a sea, which was to steal her son from her. When she heard the doorbell ring, that mother used to drop everything to welcome her boys home with a hug.

Now, it's all a memory. Now there will only be one boy coming home, kicking off his shoes and ringing the bell. How is this possible?

When the others arrive, they find me sitting there with my knees drawn tight under chin and my face burrowed into my arms.

Gösta's Birthday
February 20

Normally, we would have set the alarm clock for an early awakening but we had never managed to set it early enough. Gösta had always seemed to manage to preempt us and was always awake before us. He would have been lying in his bed listening, waiting.

Today, there was no need to set the alarm. Not this year. Nor next year. Never again.

But we celebrate Gösta's birthday nevertheless. We invite his best friend Isak to come over with his family for a celebration by the pool. We do a barbecue and we eat cake. We drink champagne. We say cheers and we weep. And I laugh because this cannot be for real.

Annette's Visit

The doorbell rings. Bertil goes to open it and I stay in the kitchen, wondering who it can be. In a flash I imagine that it is the police coming to deliver news of Gösta's death, or that they will say he has been found alive, in a hospital.

Then I see Annette. Annette is Norwegian. She lives eighteen floors below us. Her husband works with Bertil and I've become friendly with her. A few months earlier I had put considerable effort into trying to persuade her of the benefits of taking over the two guinea pigs we had bought on impulse for Julius and Gösta. She had seen straight through it, and we ended up giving them to an Indonesian woman who seemed thrilled at the prospect of giving them to her children.

Annette comes into our hallway, steadies herself and tries to say something. But she can't. She looks at Bertil and me and begins to stutter, but then she simply cries. I instinctively move towards her and place my arms around her. I am consoling her. I tell her it will be alright, that it isn't so bad. They will soon find Gösta. He is most likely in a hospital somewhere and hasn't been able to contact his parents. Don't worry, it'll be fine.

I say all this to her. *I* am comforting *her.*

Fury in the Kitchen

I am alone in the apartment. The others are down by the pool and I've come up to get some fruit from the fridge to take down. I enter the kitchen and reach for the fridge door handle. My attention is snagged by a note that has been on the fridge door for weeks. It is wrinkled and on it is written in pencil, with spidery letters: 'Cinnamon buns – warm 3 dl milk and 200 g butter to 37 degrees, add 50 yeast, 2 1/2 dl sugar and 12 dl flour. Mix together well and work dough until smooth.'

Gösta loves cinnamon buns. He must have found the recipe on the net and written it down on this very paper in front of me. We only baked them once.

Before I have time to open the fridge an overpowering rage ignites in me. My entire body goes numb and my head swirls. I am a caged animal. I take hold of the stool in front of me, lift it above my head and dash it with all my strength against the wall. It ricochets back towards me and I renew my grip, howling as I hurl it again, in another direction. The harder it kicks back, the more forcefully I renew my efforts. It is bouncing like a ball off every space of kitchen wall as I roar. I can't get enough of this. My frenzy is directed at this cursed stool that keeps rebounding at me! I want to conquer the bloody thing, so I grab it one last time and launch it out through the kitchen door. It soars across the room but by the time it slams into the far wall it has lost velocity so that instead of exploding into fragments, it slips pathetically down the wall and comes to rest quietly on the floor. It has surrendered.

Beside the sink are Bertil's and my champagne glasses. We were given them as a wedding present and they have the date and our names engraved on them. We drank from them a few days earlier when we were celebrating Gösta's tenth birthday. They have just survived my rampage.

I look at the fridge again, this time deliberately ignoring the recipe on the door, and open the door to take out some fruit for my children back down at the poolside.

Returning to Khao Lak

Julius' last day in Singapore. We are standing at passport control saying goodbye. Julius is going to return to Sweden with his Auntie Eva and the cousins who have been visiting. We have talked through at length with Julius whether or not he wants to stay here with us, or whether he'd rather go back to Sweden. He is certain that he wants to go back. We have been in touch with the school he is to start at and they say he's welcome to begin after the mid-term break. I understand how Julius feels, and I feel the same. It is meaningless to stay here in Singapore now.

We say our farewells and I leave Julius. This is the strangest farewell I have ever experienced. Bertil, Karin and I turn and leave the airport.

It is already the end of February and tomorrow we are to go back to Khao Lak to see what is happening with the search for Gösta. We have heard it rumoured that children who have found themselves alone after the tsunami are being kidnapped. This is our last chance to go back because Karin and I are soon due to return to Sweden. Julius had been determined that he would under no circumstances return to Thailand.

When we arrive at Pearl Village in Phuket everything has changed. Everything seems well organized and there are no longer desperate and injured people milling around. Anders from the Rescue Team greets us and gives us information about places to stay, prices and availability. We arrange a meeting the same afternoon when we can ask about what is being done and how far they have come. We will be able to meet with police and a forensic scientist, and Anders will also attend.

We feel well taken care of and in the evening we enjoy dinner together with Anders and some other tsunami victims whom we met in the afternoon. As usual, Karin finds some children to play with and she is soon frolicking around on the beach with them. It is a relief to see that she hasn't become afraid of the sea.

The next morning we pile into a pickup truck and prepare for the hour long journey back to the place where we lost our son. Anders and a policewoman are with us. Bertil whispers to me that the policewoman is the one who is often on TV doing public outreach when the police are looking for someone.

On the way we talk about the kidnapping rumours and both Anders and the policewoman, Eva, assure us that they have investigated every such case and every time it has turned out to be nothing more than rumour. Anyhow, they avow, those who are involved in trafficking aren't interested in white children because they stick out too much.

We have arrived and the driver turns onto the gravel road that leads down to the parking area by the holiday bungalow area we stayed in. He swings to the right and parks. All four doors open at the same time and we all bail out.

I look around me. Was it right here that I last saw Gösta? This could be the very spot from which I shrieked at the boys as I ran, seeing Gösta turn his head and look at me.

Bertil comes over and stands beside me and together we quietly contemplate the scene for a while and then begin walking down towards the beach. The bulldozers have flattened everything. Nothing betrays the fact that only a couple of months ago this place was a holiday resort, full of life and activity. The only thing that now distinguishes this place from a gravel pit is the remains of the swimming pool at the bottom of which a puddle of slimy green water has collected.

It strikes me that the reason we are here is because we wanted a relaxing holiday and for that, Gösta paid with his life. It feels sinister standing here. Here, where it ended.

We reach the water's edge and it too looks so different. Almost half a meter of sand has been washed up on land. We begin walking to where Happy Lagoon used to be.

When we reach it we see that the pine trees are still there. They have lost their bark a few meters up and higher up still we see a mattress tangled in the branches. Nothing remains of the flimsy buildings. All we see is five rough-hewn crosses planted in the sand. I think of the family from Chiang Mai who ran the business and wonder, are these crosses for them?

All of us wander aimlessly about, each enclosed with their own thoughts, even Karin. And then we gradually begin to weave our way back along the water's edge to where we had set out. Bertil takes out the camera then and says that we should record exactly what happened. Eva takes Karin's hand and says they will go and hunt for shells and pretty stones. Alone together, Bertil and I begin the filming. I speak to the camera, methodically and soberly as we tramp back up the path towards the parking area. It is burdensome to do this but it feels necessary. For Julius' and Karin's sake. Although Julius has refused to come here and Karin is so little, one day they will ask and then our documentation will be invaluable.

We push on along the small road until it meets the larger one.

"This must be where the wave hit us," I say.

"I guess you're right. That must be the tree that Julius clung on to. And see that jeep over there? That must be the one you were sucked under Ann."

"I don't know. I suppose it could be."

"And where the hell did Gösta go?!"

We both scan the bushes and undergrowth on the other side of the road.

"If he was dragged out into that then there is no way he could have survived."

It is Anders who says what we already know, but cannot absorb. We leave the road and step into the undergrowth. Although the ground is dry now, and the grasses are still lying flush with the earth, it is still laborious to walk here but I walk into the thick of it and stand in the middle looking about me.

I try to feel Gösta's presence. Was it here he died? Was it here that he failed to reach the surface? Was it here that he was forced to give up? Was it here that his heart stopped beating?

We head slowly back towards the car. Bertil has continued to film in order to capture as much as possible. I think he finds it helpful to have something to concentrate on. Karin and Eva are already by the car, talking to one another. When Karin catches sight of us she sprints towards us, eager to show us all her shells and stones.

On our way back from Khao Lak to Phuket we stop at a wall named Sight 1. It is a temporary memorial on which the names of victims of every nationality have been written in alphabetical order. We pass numerous countries, photos and flowers before we reach Sweden. So many people have hung up pictures of their loved ones but we have no picture of Gösta with us. It's possible that he is lying behind the wall because that is where the bodies are being examined, but we cannot go there.

Anders looks at his watch.

"We'll make it in time for the ceremony, he says and turns towards the car."

On a large tarmac area close to the sea, stands a white tent with no walls. At one end there is a table covered with a white cloth that reaches almost down to the ground. A bouquet of flowers is standing in the middle of the table and in front of the table, on this particular day, stand eleven coffins, each one draped with the Swedish flag.

A priest appears and begins singing a hymn. A little way away from him, by the coffins, people have gathered – these include not only relatives but medical staff, police, rescue workers and others who have helped with the identification of those who are now inside the coffins.

We stand to one side because this time, the ceremony is not for us. But Anders wanted us to see what happens. Soon, Gösta will be lying in such a coffin.

We watch. It is such a beautiful setting. As the evening sun dips lower, the light it casts over the sea softens to shades of peach and apricot. The water glimmers unruffled – the waves have almost stopped churning onto the sand, perhaps out of respect for the spectacle of grief that is being enacted under the white tent roof.

On our last day we stop by the elephant that stands some way up the beach near our hotel. The keeper told us how his elephant had saved his life when the tsunami came. It its wisdom the animal had broken free of her chains and begun striding up the hill away from the sea. Her keeper soon realized that he could not reason with her and he could do nothing but follow in her tracks. When they had reached a good distance from the sea they heard the deafening thunder of the wave as it snapped trees and ripped houses into its wake. When the elephant keeper later returned to the beach area, everything had been demolished.

In the afternoon we meet Jonas. He is the policeman with whom Bertil and I will have continuous contact all spring. Jonas will deliver regular reports about progress in the search for Gösta.

In the evening, we take a taxi back to the airport and then we are once again on a plane bound for Singapore. This is the second time that we leave Thailand and Gösta behind us.

PART 3
2006

Different Rooms

It is like entering different rooms. Thoughts flutter between the room of the present, the room of the past and the room of grief, where death is.

The present is the room in which I live. Here is the fact that Bertil and I have separated and I have moved again to another house further from Gösta. I live in a house that Gösta will never get to see. It has large windows that look out over a spacious garden edged with giant oaks. At the moment the garden is resting under a thick blanket of snow. The snow makes the ground appear to swell in soft waves and I can only imagine what is beneath. The majestic but naked crowns of the oaks make me eager to see them burst into life again. There are slender icicles dangling from the roof, just as they should in early March.

It is a gloomy day today and everything outside is grey and white. The trees are pencilled starkly against the sky and snow and although it looks cold outside, it is alluring. There is peace in nature today.

Indoors it is cosy and snug. I am content in my home, enclosed in my brightly colored walls. I have only been living here for one month. Bertil has stayed in the house we bought when we returned from Singapore. It is about a kilometer from mine. Each room is painted in its own hue: lime-green in the sitting room and kitchen, red in the hall, yellow in the corridor, blue in my bedroom and Julius and Karin have chosen colors for their rooms. I need to surround myself with color. The mats are busy with color and the furniture is upholstered in kaleidoscopic textiles. I like the Indian feel that my heavy wooden furniture gives; it gives a mystical mood to my home.

There is a packet of incense on the coffee table and I can smell their jasmine scent from where I am sitting. I bought that incense in Singapore – when Gösta was still alive. He loved the smell and would often want to light a stick. We used to lay out crisps and juice for the children and beer for Bertil and myself and then settle into our Thai cushions and indulge ourselves – and we had to have incense burning because then Gösta would be satisfied too. Sometimes he would snuggle himself up on my lap and say that he wanted to be little again. And then he would have to fight off Karin, who would want to sit there too and while they were locked in battle, Julius would be stretching himself out full length on the cushions that his squabbling siblings had so conveniently vacated. These sessions would often engage Bertil and myself more in diplomacy and conflict management than beer drinking; we wanted the children to savor these moments.

As I sit here now and survey all the objects that have come back with me from Asia and inhale the scent of the incense it is hard to keep myself in the present – I tumble unexpectedly back into the room of the past. It is astounding how real that room feels. I stop breathing initially because Gösta is so unequivocally present. He is so alive.

I can hear him, see him, feel his closeness. He is so close that I can smell his boyish skin. His voice is resonant, giggling or doing a take off of something from a TV game, or complaining about homework, or singing.

It is so painful to be in this room because I so badly want to feel Gösta close to me, with my five senses. It is like being unhappily in love, as though one is longing for someone who already belongs to another. I long for Gösta but he belongs to another world.

It is especially difficult to be here in this wonderful home with a garden full of trees that are perfectly shaped to entice a young boy to climb them. Gösta would have loved to live here. I can see before me how he would be off into his fantasy world where only he and Julius and their friends are allowed.

Everything that I possess is somehow meaningless when I enter the room of the past. It is simply a pretty facade, a shell. What do I need all this for? I would rather have lived in a garden shed with my three children than here with one missing.

It is like owning a flashy car of the latest design with leather seats and every conceivable gadget – but the engine is faulty, one cylinder is defunct. It splutters and coughs when I drive it and there is no power it. That is how life feels: perfect on the outside and hollow on the inside. I so wish I could exchange my glitzy vehicle for an old banger that fires on all cylinders.

But I have no choice. I have to live on and accept my fate. If the engine is broken, at least it's better to have a shiny chassis than a wreck.

Think of Julius and Karin.

I have left the room of the past now and passed through the present, where Gösta is missing and am being drawn into the room of grief. How can I grieve? How can I grieve for Gösta? I know I need to but there are no tears. Well, sometimes there are tears and sometimes acute pain. But mostly there's just a void that no volume of tears could ever fill.

And how can I cry now? I've cried before and what did those tears mean? I couldn't possibly have known then what the real meaning of words like grief or desperation.

How can I shed tears now as I used to when Gösta was still with us?

A New Identity

I'm sitting in an armchair looking at the window. It's pitch black outside and the glass is throwing back the reflection of a person in an armchair talking in the phone.

"Many times," I say into the telephone, "I've thought about my new identity and how I should view myself, and how I should relate to things that I know from before but that seem new now."

"What do you mean?"

"It's as though I have to do everything for the first time since Gösta's death: comb my hair, watch the traffic, walk across the floor. It's like being a child. There are all these things that I used to do without thinking. Now I have to do them again but consciously. They aren't automatic like they used to be. I have to actively think the thought, 'Now I'll put on water for tea', 'now I shall turn on the computer.'"

"Mm."

"It's like learning to walk again after an accident. Like having to think of how to move every muscle. It takes energy."

I must look unexceptional, sitting here. People are probably sitting in armchairs talking on the phone all over Sweden at seven o'clock on this ordinary Tuesday evening.

"Am I the mother of two or three children," I ask, "though I know that I shall always be the mother of three. But I have to learn everything from the start. I have to go through all these situations in order to find out what I feel."

"I don't quite follow you," says the voice in the phone.

"Well, think of a toddler who stands up for the first time, supports himself against a chair and then takes a first step. I have to do everything for the first time in a life without Gösta. I have to do everything for the first time as this new person I've become. But the uncertainty I feel isn't like that of a toddler. I've been forced to leave behind something I treasured and am forced against my will to begin anew.

I try to sort out my thoughts so that it will be coherent, even for me.

"And it changes all the time. I have to develop, experiment. I never know how I'm going to react in a particular situation. I may think that I'm going to crack up but then I don't, or I may think I'll cope fine and then for some reason I dissolve.

I tell my friend about the end of term festivities last summer, when Julius completed junior school. The sun was shining and the schoolyard was brimming with children who were looking forward to the holidays. Everywhere were proud, smartly dressed parents listening to their children singing all the usual Swedish summer songs. It was just as it should be.

I managed to stay put for about two minutes and then I slipped away and hid in the toilet for the rest of the event, bawling with renewed force with each song.

Quiet Sympathy, Late March 2006

Does it seem odd that I behave normally when I meet someone I know in town?

It's half past nine and I'm on my way home. I'm walking along the platform in the underground. Above ground, the winter is losing its grip on Stockholm and the ice is melting. I've just come down underground from Drottninggatan, out of the first rain of the year. This evening I feel alright. I'm in the room of the present and the room of grief is somewhere behind me. I can't be bothered to look back at it and I keep myself anchored to the here and now. It's a relief. I need to do this in order to keep going.

I fish my cell phone out of my pocket and call home to check that Julius has got home alright. He was going to cycle home in the dark and he had forgotten his helmet. As I hear his voice say he's home and safe, I notice a friend further along the platform. I say goodbye to Julius, put the phone back in my pocket, and stroll over to her. I say hello in a bright voice, asking myself whether it was too bright.

I don't know her all that well but I've already met her once since the tsunami, also on the underground. On that occasion, I gave her affirmation for the way in which she expressed her sympathy. But this time, I don't want to be hauled into the room of grief. I just want to feel like any normal person who bumps into a friend on the underground platform. I do not want to play the role of the grieving mother – sometimes I just want to forget about my handicap.

The train arrives and we sit beside each other for several stations. I talk away about the new house and mutual friends. As the stations pass by, I wonder what she thinks of the way I am behaving. Does she think I am too positive, too upbeat? She has no idea how I function. Perhaps she thinks I've come through my grief and put it behind me. That would be dreadful.

Although I want people to understand how profoundly I am grieving, I know that I cannot fake sorrow when I don't happen to be feeling it. Most people only see me for a fleeting moment and that may happen to be when I'm firmly in the room of the present. Or do I underestimate the sensitivity of others? Maybe they understand that one can appear happy all the same. I don't know. But throughout this journey with this acquaintance, we don't once mention Gösta or the tsunami and that feels good. It isn't necessary and I feel I am in control.

I feel satisfied when we part. Maybe I felt her wordless sympathy. We both knew what had happened and didn't need to talk about it. Sometimes words are superfluous.

I Am Two People

On the one hand, I am me, or I want to be me – just an ordinary person like anyone else. But on the other, I am a broken person. I can swing between these two instantaneously. I try to find the link between them, but mostly I just want to be an ordinary person.

But I carry a heavy burden and my task is to try to be an ordinary person who happens to be bearing a heavy burden. Sometimes it seems lighter, but sometimes it feels intolerably heavy. Underneath it all, I am a positive person and it doesn't suit me to be melancholic. I don't want to, but I am unable to prevent myself from being sucked into the room of grief even though I loathe being there.

I live in two parallel worlds.

And yet I am captured in four dimensions-time and space.

I want to escape.

Is Gösta out there? Beyond these dimensions?

No, I don't fear death.

Since Gösta is there, how could I fear death?

Gösta is simply in Another Place.

What does it matter that I was close to death too?

It is only worth living if one can live fully.

There is no point if simply existing.

Why must we live, no matter what?

Think Julius and Karin

Looking Back: The Deer, Spring 2003

It was Friday afternoon at last. The week was over and I was in the car on the highway, en route to pick up Karin from day care. Julius and Gösta were already home. The snow had melted and the sun was on its way down in an eggshell sky. Despite the chill on the air outside, the sun had heated up the car enough to make it smell of car, with all the winter dirt from muddy children in waterproofs.

I had just overtaken a car and was in the left lane when I noticed the deer on the parkway in the middle of the highway. I only had time to wonder if I should swerve to the right or the left as it leapt directly in front of me. In a flash, I envisioned it continuing over the road, no, stopping and turning back when it saw the car.

It was all over in a moment, the impact so sudden that the car reeled and zigzagged back and forth a few times before I was able to straighten it up and steer it out towards the shoulder. I stopped and got out. One headlight was smashed in and the housing was hanging by a thread like a gouged-out eye. I surveyed my surroundings but could see no trace of the deer so I got back in the car and continued warily on my way to collect Karin. When we got home, I rang the police and reported the accident.

Later that evening, after I had spoken to the hunter who had located the deer and assured me it had died immediately, I found myself crying about what had happened. I asked myself what I was crying over.

It seemed so tragic to have killed an animal that would have been free to roam the forests if it hadn't been for me. I felt guilty about having extinguished a life. But then, I reasoned, who else apart from me is bothered about this deer? The deer itself did not suffer, probably not even aware it was dead. It was apparently one year old. It would already have been rejected by its mother who almost certainly had a new fawn by now. So it would not be missed by her. Its father would not even know that it had offspring. Siblings, cousins, grandparents? Hardly. Companions? Unlikely.

This young deer was most probably entirely alone and would have had no one who missed or grieved for him. So why should I be upset? Everything had happened relatively quickly so it probably had no concept of what happened. I tried to understand why it was so horrifying that a living creature had to die, even though it was just an animal.

In the natural world, death is an entirely natural event. Everything dies: plants, animals and people. This is the natural order. So why lament it?

After reasoning with myself for a while I felt considerably better. I decided that there is nothing to pity. The deer required no pity since it was dead – it no longer *was*. There was no one who deserved pity because they were grieving for the deer. In the end there was only me to pity because I had caused all this, but I would only have deserved pity if there was someone who was sad because the deer had died. But there wasn't, so there was no point in pitying myself either.

Spring

I sometimes wander around in Täby Centrum. It can feel good to mingle with ordinary people. I try to imagine how it might have been, how I might have strolled from shop to shop and then bumped into Gösta with a gang of friends. Gösta never reached the age at which he would be embarrassed by his mother. On the contrary, when I used to come to his school he would drop whatever he was holding and bound towards me with open arms calling 'Mamma!' He would have done that in Täby Centrum too.

On this day, I planned to buy some large items from a shop called Indiska. I had taken the car because it would be too much to carry. I found a parking spot close to the entrance and climbed out of the car into bright sunlight. The mercury had dropped to minus ten degrees in the night but now the ice was forming pools on the asphalt and the air was becoming moist.

It was early afternoon and midweek so there were not many customers in the shop. Being such an incorrigible fan of their vivid textiles and ceramics, I decided to have a scout through that section before going to get the item, a big table. I thought I might find some small, unusual gifts to take to people instead of flowers when I was invited over. I closed my eyes and sniffed the dark wooden ornaments and it took me back to India, where we had been a year earlier, the whole family.

And then I heard the song, a song by Enya floating through the shop from the loudspeakers. I had no time to ready myself – I was trapped, cornered with no chance of escape.

My whole body remembered and a tide of emotion swept through me. I wanted to get out of the shop and sit down in peace but I was captivated by the music that I used to listen to regularly when I was expecting Gösta. I wanted to remember just that but nothing else. Nothing more.

But I was in a hurry and must finish shopping. I was frozen, blocking the path of an irritated mother who was trying to navigate her way past me with her pram. I stepped sideways to let her pass and swallowed hard. With the music in my ears and a constricting throat, I strode determinedly towards the cash desk to ask the assistant for the item I had ordered. She was keen to help me carry it and together we shuffled towards the shop entrance. And then they came – the tears. I decided to tell the bewildered shop assistant why I had to cry and she too began to sob. We put down the package and she came to me and held me tight.

"There are no words," she said and held onto me for a long time.

Looking Back, Newborn Gösta

I can see Julius the first time he saw Gösta.

I am sitting on the edge of the bed at Danderyd Hospital. Julius is in front of me. He is two years and three weeks old. In front of him lies Gösta, who is twenty-four hours old. Julius is marvelling. He inspects Gösta and then strokes his head gently and I believe that this is the beginning of a lifelong friendship. I hand Gösta to Julius to hold.

"Here," I say. "This is your brother for life. I hope you'll have lots of fun together and that you'll always have each other. You will grow up together and share so many things. You'll be together."

I begin imagining how Julius and Gösta will play together, share secrets, grow up and cement their own ideas – ideas they will test on us. I imagine them growing from boys into men.

Julius bows forward, delivers a kiss to Gösta's forehead and says, "He's cute."

How Is Julius?

I look at the clock and see that it is just after two o'clock. It's a weekday and this is my week with the children. Julius will be home in about an hour and I am sitting waiting with a cup of lukewarm coffee in front of me. I begin thinking about my communication with Julius. We rarely talk about Gösta. I don't know what to say to him. The disappointment about things not turning out as I'd imagined is piercing – Julius will not grow up together with his brother.

But Gösta is ever-present in my thoughts.

I don't know how to approach Julius. He lives so much in the present. He is doing well at school and has plenty of friends. He seems to be doing fine. From the outside, he appears to be coping well with the situation. But then, I wonder, what does one look like if one isn't coping? I am anxious about the fact that Julius is coping so well.

Last summer I received a letter from a woman who lost her younger brother when she was Julius' age. She wished that her parents had made her talk about it, made her express the fact that she was finding it so hard to cope. Maybe Julius feels differently but I worry nevertheless about what may be concealed behind his seeming ability to cope so well with the death of his brother.

But I understand Julius as well. I behave much the same. I also want to live in the here and now. It's just that I find myself in situations that I can't handle, when I have to confront the fact that Gösta is dead and can't keep it at bay. It just overwhelms me.

With Julius' permission, I have spoken to a child psychologist. I spoke in my capacity as mother. This was supposed to be about him, not me. Why then couldn't I keep a lid on it and stop myself from crying? I know that I am able to talk about all this without weeping – what is it that makes me crack?

Maybe it is just so close to the surface that it needs almost nothing to unleash it. Maybe it was because the psychologist explained that losing a child is the hardest thing of all – and that is what I have done. Maybe it's because of this added dimension of grief – the fact that I must watch Gösta's siblings grieve for their brother. Seeing their pain is so painful.

I explain to the psychologist that I am fearful of the fact that Julius carries so much inside himself and that it is becoming burdensome for him.

Ann Hageus

The Puppy

Julius wants to get a dog. How can I possibly say no? It's the least I can do for him. He has been leafing through books about dogs and after lengthy deliberations has decided that he wants a Shetland sheepdog.

When Julius is away skiing with Karin and Bertil I call up a breeder. It turns out that she has a litter of pups that will be ready to leave their mother in one week's time. I decide there and then. The breeder even lives just outside Enköping, not far from where Gösta is buried. Julius is not to know anything. It will be a surprise.

A few days later they return from their skiing trip. The children are to stay with me now for a week. At the weekend, I ask if they would like to come to Enköping. They think nothing of it since we quite often go up there.

On the Saturday morning we all get into the car and set off. The journey goes quickly and when we continue past the exit to Villberga I see them looking at me, but I say nothing and continue driving, chuckling quietly for myself. This will be fun. Neither of them suspects anything. When they begin asking where we are going I tell them that we're just going to look at something first. We arrive at the school I worked at before we left for Singapore and I point out a few landmarks. A little further along the road I find the house where the breeder lives.

We ring her doorbell and hear a chorus of barking from within. The children look confounded but say nothing. Eva opens the door and welcomes us in. She sounds just as she did when I spoke to her a week earlier. She leads us into her sitting room and squats down on the floor. Four small pom-poms scurry towards her.

"So," asks Eva. "Which one do you think is cutest?"

Julius is already on the floor and cradling one of the pups that has claimed him. Julius is speechless as he watches the muddle of pups then scramble around the floor and he reaches down to gather up the one that jumped up onto him before. He lifts it into the air and then brings it to his face, stroking the silky fur against his cheek before he places the pup on his lap again.

"This one," he says slowly. "This one is the cutest."

Maybe he is beginning to understand.

"Julius," I say. "That one is now your dog."

He looks at me, puzzled. He is trying to take in what I have said.

"We've come here to buy a puppy, a Shetland sheepdog," I explain.

Julius' eyes are like saucers. He looks at me and his mouth drops open but he can't get a word out. Then he looks down at the little creature in his lap and it is looking back at him now with twinkling chocolate-colored eyes. Julius is trembling, his body saying what his mouth cannot.

Karin has been playing with the other pups but listening at the same time and when she hears what I say she bounces up and squeals with delight.

"Mamma! Are we really going to get a dog?"

"Yes," I say. "We are. I've been thinking about this for a long time and I reckon it would do us good to have one."

An hour later, we are sitting in the car again but now there are four of us and one has his small damp muzzle pressed against the window.

I had no idea at this stage what I had let myself in for; I didn't realize that owning a puppy is quite different from owning a dog. The months that followed were a constant trial with house training, mangled shoes, filthy dog paws all over the beds. But the children were so thrilled with Kasper, as he came to be known, that it seemed worth the trouble.

***

Margins: it's okay to walk along a broad path but if it becomes narrow there is no margin. If anything goes against me I feel rage first boiling in my stomach and then, within seconds, it has overflowed and saturated my entire body.

***

I don't want to feel bored, but am I able to feel bored, even if I do want to? The problem is that what used to feel like boredom in a former life inevitably becomes the room of grief now. And so I don't want to feel bored. I have to keep myself on the surface.

Why Is It So Terrible When Someone Dies?

This morning Bertil called from Falun to tell me that his father, Ville, had passed away. It was not unexpected. He was old, tired and sick and Bertil had long been waiting for this.

But however well we know that we must all die, every time it happens it is as though the world stops spinning. We lose our bearings. Why?

How did I feel when I heard about Ville? His death didn't compare with Gösta's, but I was affected by it. It was something like a vacuum, a kind of stillness and instant of something outside of the everyday reality I live in. What is this about?

Maybe I am wondering whether Ville is now together with Gösta. Where is Ville now? He was here so recently – I could have rung him, asked how he was and heard his own voice reply. He had a place in our world. He still has, though only as a body, an empty vessel. The space he will occupy from now on is so radically different.

There were facts and truths about Ville – how he was doing, what he thought, what he did and was going to do, what he said. One could reach him, prompt a reaction in him. But now all that is gone. How does this happen? Here one moment and the next, so indescribably far away.

Strange. Irrevocable.

The facts and truths about Ville remain, but only within those of us who still have a place here. He can no longer communicate his existence and is dependent upon us for his continuance.

***

My miscalculations.

It is onerous to hold myself up by the arms. I have to be so egotistic these days, partly for my own sake but also for Julius' and Karin's sakes. It is essential because they can only fare well if I do.

Looking Back: India, November 2004

Gösta! God, how I long for you! Reclaim my memories.

I have been deprived of my memories from before 26 December 2004. When I think of some memory from Singapore or our travels while we lived there, it is so tormenting that I prefer not to think of them. Shit! I had so looked forward to gloating over those memories.

Now I can't be damned with those cursed memories of magical times with Gösta. I hate them.

We had dreamt for ages about going to India – imagining that vast, vibrant country with its heat and spices and curries. For several years we have been supporting a child in India through the International Children's Fund and we have been writing to her regularly. Shridevi was only six years old when we were first put in contact with her. At the time of writing, she is already fifteen. When she was little and couldn't write, she used to send us pictures she had drawn. They were so different from the pictures that Swedish children draw, full of flowers in dazzling colors. And later she began to write.

Her letters are also works of art. She decorates the paper with swirls and flourishes and the words are written in her native tongue, Kanada. A translation always accompanies her letters so we can read about her life, her school and friends and what she enjoys doing at school or in her free time. In one letter she told us that she had a new baby brother and that meant she has one elder sister and two younger brothers.

For years I nurtured a hope that we would one day be able to visit Shridevi in India and when we moved to Singapore it was as though my hope might become a reality. I realized we might never get the chance again. We had been living in Singapore for a year and three months when we took off for Bangalore, unaware of the fact that this would be our last holiday together.

My memories of that trip are intense and happy. Everything went well and no one got sick, nothing was stolen and the children enjoyed themselves.

In the run up to the trip, I had been concerned about the children. Although we had heard many fantastic stories about India, we also knew about the poverty and deprivation. Gösta was easily upset by things like this and found it particularly disturbing when he witnessed children suffering. In his world, all children were as well cared for as he was and he found it hard to accept it when he began to discover that this was not the case everywhere. I explained that children who were materially poor may not be any less happy and that they often had large families and a strong sense of belonging.

I was most worried about the prospect of one of the children being kidnapped. Just before we left for India I could work myself into a panic conjuring up images of one of them being dragged into a car that took off at full speed into the traffic. I promised myself that I wouldn't let them out of my sight and that I would try to keep hold of their hands as much as possible.

My apprehensions proved to be unfounded. The people in India were so friendly and there must be photos of all of us in hundreds of Indian photo albums by now.

We spent hours ambling through the parks in Bangalore and soon caught the attention of Indians who wanted to take our picture. The first time we were asked, we were astonished – no one would do this in Sweden. We gladly posed and smiled but we only managed to walk a few more meters before we were asked again. It was not only families who asked, but single men, old ladies and all kinds of people.

It was exciting to be so popular, but it was our blond children whom people were really intrigued by. It was Karin who attracted the most attention and, unsurprisingly, she who was the first to tire of it. By the end of her trip, when her cheeks had become dotted with bruises from all the pinching, she flatly refused to pose any more for the cameras and growled and snarled at people instead.

We have one photo of us sitting on the grass in a park surrounded by some twenty Indians, all staring in enthralment. In India no one is shy about gawping at those who look different.

We travelled mainly by bus and train. On our way to Shridevi's village, we crammed ourselves into the third class section of the train, and since there was so little space on the wooden benches, the children climbed up onto the overhead rack and we sat underneath them next to an Indian family. We tried to strike up a conversation with them, but when we ran out of common vocabulary, their daughter began to sing instead. She sang us an Indian folk song in long, clear tones and then we were all pulling out food parcels to share.

And then I noticed that Bertil, Julius and Gösta had disappeared. I asked the Indian family to keep an eye on Karin while I went to look for them and began walking backwards along the aisle. All the windows were open and warm air was flushing through the carriages. I found Bertil standing at the far end of the train by an open door. I looked down and saw that Julius and Gösta are sitting on the floor beside him with their legs dangling out through the door opening. They weren't holding onto anything and I realized that it would only need for the train to jolt hard and they would be flung out. My heart was hammering fiercely against my ribs as I grabbed each bewildered-looking boy by the arm and yanked them away from danger.

Towards evening we had to change trains and board a night train. It was wondrous to lie there listening to the rhythmic clanking of the wheels and to peek out through the windows at each station and watch the crowds milling about on the platform.

When we reached our destination the next morning, we settled for Indian breakfast at a cafe not far from the station. We were served steamed rice cakes known as itly and naan bread together with a range of sauces to dip these in, and the children were aghast that they were allowed to eat everything with their fingers.

We were due to meet up with Shridevi that same afternoon. It was hard to believe that what had for so long been just a fantasy was now a reality. It is so easy to wish and dream, but how often do we live out our fantasies? It is then that one wants to stop time and remain in the moment as long as possible.

The road we travelled on was lined with goats and oxen, men in white robes and turbans, women in flamboyant saris. When we reached the village and began to make our way towards Shridevi's house, we became a magnet for more and more village children and even adults, so that we were far from the only ones who stood waiting on the threshold for Shridevi to appear. When her door opened, not only Shridevi but her whole family came out to bless us with incense, flower garlands for our necks and vermillion for our foreheads.

I was surprised by how moved I felt. It was now that I realized how much this all meant not only for Shridevi and her family but for me.

We were shown into the house and offered chairs. Half the village followed us in to the small room. Everyone was curious about us and in particular about the children's blond hair. We presented the family with the gifts we had brought and chatted a little with the help of our project leader, who acted as interpreter. Then a young girl from the village sang us a traditional song. Julius and Gösta were clearly embarrassed by all the fuss. When we were shown around the village all the local boys tried to get close to them or hold onto them, while the girls took turns to carry little Karin.

We spent the next few days with Shridevi and her little brother. The children were shy at first but after a day or so they began to thaw and although they had no common language they managed to communicate.

Our project leader, Mr. Suddhash, showed us around the area. We visited the 17th century mausoleum Gol Gumbad - the burial site of Muhammad Adil, his two wives, his lover, his daughter and his grandchildren which has the second largest dome in the world, surpassed only by St Peter's Basilica in Rome.

We sat down to rest in front of the building, but the children couldn't sit still. Julius and Gösta challenged Shridevi's brother Vital to a game of tag but although he was younger, he easily outstripped our boys.

On our last day in the village, I was given a sari by the village women. It was petrol blue with threads of red and gold woven into it. The women helped me crease and drape the six meters of cloth correctly over my body until I looked almost authentically Indian.

It was these singular memories of Shridevi, her family and her world that I thought I would treasure for ever. But should I love or hate these memories now?

We arrive home in Singapore on 19 November 2004. Gösta has thirty-seven days left to live. In those thirty-seven days I failed write to Shridevi to tell her how much we had enjoyed our stay with her. I intended to say that I hope this will mark the start of something new, that we will meet again, that the children will remember our trip for ever and that it may inspire them to visit India again in the future. Nor did I get around to sending her the pictures we took of her and her brother smiling into the camera together with our children.

Many months later, I managed to pen a letter to Mr. Suddhash explaining what had happened and telling him that I couldn't bring myself to write to Shridevi, though I wanted her to know how much we had enjoyed meeting her and her family. I asked him to convey all this to her for me.

A couple of weeks later, I received a thoughtful reply together with photographs of a ceremony they had held in honour of Gösta. They had placed a large photograph of him and all around it lay flowers. The family and all of those we had met from the organization were standing in a ring around the table praying for Gösta. It was this that then gave me the strength to later write directly to Shridevi.

The Airport

I am standing in the domestic flights area of Stockholm's Arlanda airport, en route to Malmö to attend a board meeting with the International Children's Fund. The hall is huge with a rounded roof so I can see all around me as I take the escalators to the upper floor.

It hurts to be here. Almost two years earlier I was going down on this escalator with Gösta's hand in mine having just met him off a flight back from a visit to his friend. He had been carrying a large poster of a monster truck under one arm. It is bizarre to think that this is the same hall, the same escalator – so similar and yet so different.

I have a window seat on the plane. We have already climbed above the clouds and the sky is clear now. I lean against the window and close my eyes.

I can see Gösta's face and I hold onto the vision. I involuntarily reach out with my finger and touch his forehead. I feel the even pulse in his temple and when I reach his cheek I cup it with my whole hand and continue stroking down towards his chin. I bore the tip of my finger into the dimple in the middle of it and Gösta laughs. Then I run my finger along the soft curve of his lips, up the line to his nostrils, over the bridge of his small, straight nose. When Gösta was one year old he had three freckles on it. The following year, he had seven. Now his whole nose is freckled and even his cheeks. I love those freckles. My finger meanders up over the nose and then follows a pale, sleek eyebrow, down over the eyelid. Gösta closes his eye. I stroke as carefully as I can because the lid is so delicate and thin.

Gösta is so perfect. I withdraw my hand and look at his face, so lovely.

Late May 2006

On a bright spring day in late May, I take a walk together with my sister My and the children down towards the Museum of Ethnography. Our dog Kasper is with us. We keep looking over our shoulders to see if the bus we want to catch is coming yet. Karin and her cousin Andrea take turns to hold Kasper's lead and the puppy bounces along between them, picking up sticks here and there. Julius and Olivia are walking a way ahead of us and they suddenly call out that the bus is coming and they wave their arms furiously for it to stop. The rest of us come panting after, dog, sticks and kids all piling on at once. We plonk ourselves down in the middle of the bus and it feels as though our noisy crew has filled the whole vehicle. I turn to the woman sitting beside me and she smiles at me, and I smile back apologetically. But then I recognize her face. It is Helena, the daughter of one of my parents' closest friends. It must be twenty years since I saw her last.

When she sees who I am, her face lights up for a second and then crumples. She doesn't know what to say but it is obvious that she knows. Without saying anything she somehow conveys empathy and there is no barrier between us. I assist her by saying that life is so tough. Helena agrees that it is too horrible and then the conversation flows naturally onto other topics. She is going to meet her daughter at the Maritime Museum and we chat until it is time for her to get off. Although we hardly know one another and seldom met, I am left with a feeling of mutual understanding when she is gone.

One and a half years have passed and there are still acquaintances whom I haven't met since Gösta's death. I am sitting on a terrace out in the countryside remembering that chance meeting with Helena and thinking of how differently meetings like that can turn out. The way a person greets me now can impact powerfully on how I feel and the meeting with Helena showed that there are people who are not afraid to confront pain and who one way or another avoid closing it off.

The worst is bumping into someone I know and then finding that they make no attempt to indicate that they know but just pretend everything is the same as always. This creates an impenetrable barrier between us that will remain as long as the other person continues the pretend. I cannot help others to dismantle that barrier because I have enough to deal with without having to shoulder responsibility for closing the gap between us. Instead, I choose to avoid those who I feel aren't able to give me the acknowledgement I need.

I am sorry, but at the moment it is I who need support, understanding and recognition. I cannot pay heed to those who can't bear to meet me as I am: a woman who has lost one of the things that she values most, her child.

I am sorry that I make such demands and seem quick to judge.

I am sorry that I don't help others but instead demand that they help me.

I am sorry that I now put my feelings first.

I am sorry if that means that I may lose some acquaintances for ever.

Ljungbyhed, Pentecost

I am sitting by a round table with Karin and some friends. One of them has her mother with her. We have just stocked up from the buffet with Serrano ham, Italian sausage, olives, salmon and vegetables and the adults have charged their glasses with fine red wine.

The atmosphere is jovial in here. We are in a large room with some fifteen similar tables, each crowded with dinners like us chatting or chortling. Some people already know each other, but not all. We have spent the day together participating in various activities. The children have been training circus tricks and testing model aircrafts. We have been painting large round stones or self-portraits and in the evening there will be a karaoke and a bar. Karin is in her element. She has already entertained us all with her Hackebacke song, a string of dirty words, and she has made many new friends.

I take a sip of wine and look out over the gathering. A man at the far side of the room catches my attention – I know that he lost a son, a nine year old. My gaze settles then on a woman at another table. She lost two small children who hadn't even started school. Behind her I can see a couple of young women in their early twenties who both lost siblings. At the table next to ours is a woman who lost her mother and at the next table, a man and then a woman, both of whom lost their entire families.

At my table we can count the loss of four children: Gösta, Amanda, Joel and Linn. The jollity may seem deceptive, as though it is stifling the collective grief here, but I don't think it is a deception. People have travelled far to attend this meeting of that the organization "Vi Som Finns" has arranged so that the tsunami survivors can see each other again. Grief and joy can go hand in hand and no one needs to hide their feelings.

In the evening, one woman tells her story as the karaoke starts up over in the bar area. As she begins, the noise from all the activity fades into the background. Her story is like so many others. At first, she noticed that the sea looked strange but it didn't seem dangerous but then she realized, too late. She had taken hold of her young son but the water wrested him from her arms, dragging her down as well until she was pinned under some wreckage. She surfaced moments later and took the breath that was to save her from death. She looked desperately about for her family but was drawn out into the sea by the next wave.

The part of her story that tore at my heart was when she described how she found her daughter, who she believed dead, at a collection point. Her husband, whom she had only just found, walked into a room there and let out a loud cry. She is still numbed by the feeling of being neither alive nor dead but she follows her husband and sees a stretcher. On it lies her daughter. She moves towards the stretcher, lies down beside the little girl, puts her face beside her daughter's and begins to stroke her child's cheeks. She tells us that she could feel nothing – she is unable now to say whether she felt sadness or joy. She was simply there with her child and there was space for nothing else.

As I listen and tears are falling on my cheeks, I feel as though I am being transported back to that chaos, where I was beyond having feelings, beyond caring about anything, back to the place at which I was numb to the feeling that Gösta might be dead, just as this woman was numb to the fact that her daughter was alive.

That night I couldn't sleep. Karin is curled up asleep in the bunk bed beside me. We have a room to ourselves that has sleeping space for twelve: six bunk beds are lined up against the naked walls. In the middle of the room is a metal cupboard with a padlock on each unit. Karin has made use of all the space by putting her clothes everywhere and putting some of the bottle caps she's been collecting during the evening onto each bed.

In the window that looks out over a large field is a collage of photos of Gösta. I am lying in bed looking up at him. He is smiling and pulling different faces. In one picture he is making the sign of triumph and in another he is cycling as fast as he can.

I think of a story that I heard earlier in the evening. A woman about my age told me how she had tried to rescue a little girl of about eight. The little girl had injured her leg and you could see the bone poking through the torn flesh. But then a new wave had washed over them and they were separated. The woman never saw the child again.

I don't know how Gösta died. I don't know whether he was injured and almost survived or if someone almost managed to rescue him. That little girl could just as well have been Gösta. It is so dreadful to look at those photos of him and know that we will never be sure about what happened to him.

***

Grateful for life?

There are those who believe that surviving the tsunami has given them a second chance and that we should be grateful that we are alive. Why? To whom should we show gratitude? I did not ask to be born and I definitely did not ask to survive.

Maybe Julius and Karin can be grateful that I survived and that they did not lose a mother. But I have nothing to be grateful for except for the sake of my surviving children.

Gösta lived his whole life.

I lived my whole life until Gösta died. What I mean is that with him, a quarter of my life died. I shall try to live fully with the three quarters that remain.

Midsummer 2006

This will be a peculiar midsummer, unlike any I've celebrated before. I have decided to say no to the invitations I have received but it has taken me quite some time of uncertainty before deciding that I want to be alone.

I shall celebrate midsummer with Kasper, our dog. I look forward to a quiet weekend without any demands made of me. Julius and Karin have gone to Falun with Bertil. This is the first time in ten years that I am not going with them and it is the first midsummer since the children were born that I shall spend without them.

It feels wrong just now to meet up with people for parties. Above all, I don't want to meet people I don't already know. I don't know what role I will have or want to play.

Midsummer Eve looms but right now, the sun is shining and I can take a cup of tea and a sandwich out into the garden to enjoy. I position my red wicker chair by the wall where the warmth has already chased away the morning chill. When I sit down, Kasper lies down at my feet. When a dog rests, it radiates restfulness with its whole body. Kasper exudes contentment all the way from the tip of his nose down to the tip of his tail. The only movement is a light breeze that plays with the tones of my Indonesian wind chimes, clonk, clink, clonk.

Now I can see what has been hidden under the snow all winter. There is bare rock, some two hundred and fifty million years old – it was here long before humans. It was here when the Earth simply *was*. It pre-existed values, questions, philosophy, existential quandaries, ideas of the future, grief and loss, meaning and memory. These, I believe, are the things that distinguish us from animals.

The rock simply exists. It will continue to exist until something happens that destroys it. It doesn't care. It doesn't care because it cannot care. It is just a rock. If it has any value that is because I award it value. And I have awarded it value. I find it graceful the way it slopes down towards me. There are cracks in the rock that criss-cross the veins. A poet might liken it to a human hand covered in bluish blood vessels. But I don't bother with such meaning making; there are naked, crass explanations for why those rock veins are there.

The rock has a natural trough that I have filled with water for the birds. I've created a personal meaning; it's for the birds. If I hadn't done this, then the rain would have. But the rain would have had no intention; it would simply have happened. The rain simply *is*. It doesn't do anything for the sake of people. There is a naked, crass explanation for why rain falls. It is I who award it meaning, such as enabling the birds to take a bath.

I sit and watch the birds arrive and splash about in the bath that I've created for them. It's a pleasure to see them. I transfer my own feelings to the birds and imagine that they are basking in the warmth of a summer morning as they perform their ablutions.

There are many different kinds of birds: magpies, blackbirds, wagtails, doves. I can pretend that one of those birds is Gösta. If one of the doves had been white, I might have believed it was he who had been sent down from Nangiala to let me know he was waiting for me there. I don't need to worry about whether the birds are actually enjoying their bath or not. Maybe they only bathe instinctively to rid themselves of the mites that would otherwise itch and ruin their plumage.

The meaning I find right now is to experience peace and proximity to nature – nature that simply *is* and which acquires the meaning I ascribe to it.

I get up and walk over to one of the flower pots that I've placed around the garden. I like to watch the plants growing. It's thanks to me that they can grow just here. I've planted bulbs and seeds and have watered and cared for them. My reward is to get to see them flourish. Are the plants grateful? I doubt it. They simply *are*. They don't care. It is I who ascribe meaning to them, for my own sake. And I feel good when I see them growing. There is no meaning in this for anyone except me.

I pick up my yellow watering can and fill it from the tap. As I water the plants, my nostrils are filled with the scent of thyme, basil, mint and coriander. Each scent bears a memory. When I have finished watering, I put down the watering can and take a walk around the garden.

The oaks are in full splendour and as the branches softly rustle, the tones of green in the foliage shift. Further down, the grass is getting long and it is highlighted with wild flowers. The snow has been hiding a dormant meadow that is now bursting with clover, cow parsley, buttercups and the smell of earth and dry grass. I love my garden. It lives its own life and I want to make as little impact on it as possible. Here, I too can just *be*, just *be*. I sit myself down in the midst of the long grass and look about me, waiting for something – or nothing – to happen.

I look up and see some tufts of cloud that are about to eclipse the sun but at the moment the brightness is still making the greens of the garden gleam sharply against the sky overhead.

I can look up at the heavens but not see into them. I don't want to see Gösta there. He isn't there. If he is anywhere, it is inside me. In order to find him I need to look within myself. He isn't out there in the shape of a bird or looking at me through a hole in the sky. If humans have a soul, then I now have two.

But I find it just as difficult to look into the heavens as it is to look at myself in the mirror. It took a long time after Gösta's death for me to be able to look at my image in a mirror. Now, when I look in a mirror, I only look superficially. I can't bear to meet my own eyes because I don't want to look into the eyes of Ann who has lost her son. I can't accept that this is me and I can't come to terms with the fact that this is my lot in life.

The clouds are huddling together and the wind is picking up. The branches are crackling and the smell suggests that rain is on the way. There are goose bumps all down my arms now so I get up to go inside. Kasper and I just manage to get ourselves and my wicker chair inside when the first heavy raindrops come.

I stay by the window and watch. It's hard to believe that it's the same garden now as a few moments earlier. All the shadows have melted away and wind has blown away the sharpness of the colors. The heavens open suddenly and release sheets of rain. So, all my plant-watering and birdbath efforts were a waste of time. But that's how nature works. It has no guilty conscience just because it made a mockery of my endeavours. It simply *is*.

The house has become so gloomy that I search around for the matches and then light all the candles I can find. I pretend I am nature and that I don't care in the least about the fact that it's raining at midsummer but, unlike nature, I am thinking. I am thinking that it doesn't matter that it rains, and I even think it's cosy. The glow of the candles chases away the rawness of the rain.

I sit down on the sofa and turn on the computer. I surf about for a while and then settle on the Swedish TV website where I notice the word 'tsunami'. Curious, I click on it and find links to the first news reports that came through on 26 December 2004. There were six special reports and four regular ones that day. I look through all of them.

Now, almost exactly one and a half years later I can sit here looking through what was said on the news. The reports described how an earthquake had taken place in Southeast Asia. There were no reports yet of deaths or casualties. Then it became clear that there had been deaths. There were one thousand in Sri Lanka, as many in India, around five hundred in Indonesia and three hundred in Thailand. On the nine o'clock news it was confirmed that two Swedes were among the dead, figures for which had now reached estimates of twelve thousand. Travel agencies were advising tourists not to travel to the region and informing those who already had bookings that they would be reimbursed. A couple was interviewed at Stockholm's Arlanda airport and they said they would travel anyhow because they didn't think it was as bad as reports were claiming.

Five hundred and forty-three Swedish people, more than two hundred thousand deaths in total.

I close down the computer and the memories come flowing back. I take them with me on a long walk in the forest with Kasper. As I walk, I can hear the distant sounds of people celebrating midsummer and I'm thankful I am not involved. I know that Julius and Karin will be alright with Bertil. So I can just *be*.

All the questions I have about what happened make me feel so powerless, since I shall never get an answer. So sometimes it feels good to just let go of them and ask myself instead, 'What shall I do next?' And then do it. A bit like Kasper. He doesn't seem to make plans, he just lives. If he's tired, he goes and lies down, and if he wants to play, he comes and tugs at my trouser leg. That's how this weekend has turned out for me too and maybe I shall continue to live like this.

***

I have begun to relate to people differently from before. Last year, when everything was so chaotic, I was outgoing and sociable. I never said no to an invitation to a party or a dinner. I think I was on autopilot. I suffered from a kind of pretentiousness and I applied for all kinds of jobs that were well beyond my level of competence. I seemed to have lost all my reference points and my perspective. On the Monday after Gösta's funeral, I attended a job interview. It was insane.

But I can't do this anymore. I feel like a cartoon character that jumps off a cliff and then discovers that she's in mid-air, but keeps on running on nothing for a while and then all at once drops to earth. It was easier last year as well when everyone in Sweden was still conscious of the tsunami. It was tangible everywhere. I could tell people what had happened and show my sorrow about Gösta's death and people were tuned in and able to empathize.

Now, everyone has moved on. There have been new catastrophes, scandals and horrors to occupy people's thoughts while I carry on living with the tsunami. I have stopped on the same spot and it is now so much harder to drag other people back into it. I am afraid of how they will react. Sometimes I have told people what happened and received no response at all, as though I had just told them something every day and ordinary. That is the worst.

I have been a dinner parties where there are people I don't know and not known what role to play. Maybe they don't know anything about my story. Should I keep up a pretence, wait for a question that will give me a chance to reveal the truth – such as if they ask how many children I have – or should I take the initiative to say something?

This is becoming laborious.

There are brave individuals who dare to step right into the horror and stay there. Just a couple of words can be enough to open a channel.

To feel fine. We ask about this whenever we meet people, 'How are you doing?' We're expected to reply that we're fine. But what does that mean? Even under normal circumstances we are expected to say that we're fine even if we aren't. No one wants to listen to a diatribe on why we aren't feeling fine.

Nowadays I cannot bring myself to say I feel fine. I don't. I don't have the basic requirements for feeling fine. And what are they? Well, I guess they are that one's closest family are alive and healthy and that one is healthy oneself and that one has a source of livelihood ... and after that the list continues in diminishing significance. So no, don't ask me to say that I feel fine, or even that I feel better.

The easiest thing is to meet people with whom I've had regular contact throughout the year, people who have accompanied me in my grief, especially those who were affected by the tsunami. I regret that we are scattered all over the country and wish it was easier to meet up.

***

I look at Kasper and he wags his tail. He's lying on my bed and it warms my heart to see him. He has followed me into the bedroom, watched me put on the computer and sit down and then hopped up onto the bed to keep me company. It's comforting to feel warmth for a dog. He is so loyal and affectionate and he looks so secure curled up on my bedspread – completely assured that he will be well taken care of. But this has nothing to do with Gösta. It isn't because Gösta died that I feel warmth towards a dog. My affection for the dog and my grief over Gösta are entirely distinct. Maybe this is what it means to be alive – the ability to feel happiness when we see another being thriving.

Lacalm, Summer

We have been invited to our neighbours for lunch at their beautiful house out on the countryside not far from Toulouse in France. They have a stunning view over the neighborhood and we are to eat on their spacious verandah where a table is already laid. The sun is hot and the sky is brilliant today. I am sitting by the aperitif table with a glass of chilled rosé wine in front of me. Around me are the members of the family who have invited us and my father.

Julius and Karin are strewn, half-lying in the shade on the cushions of a bench a little way from me. I watch how Julius bends over his sister and begins fiddling with her hair. He takes a strand at a time and rearranges them so as to straighten up her parting. He is so meticulous and purposeful. He looks gentle and his movements are loving. Karin allows him to fiddle; she seems to like it and looks a little proud to have a big brother to take care of her. It is rare to see them showing affection for one another. They are mostly irritated with each other and squabbling so it is gratifying to see them like this.

But my pleasure is broken by a stab in my heart that spreads up to my throat and towards my eyes where tears begin to sting. I reflexively try to hold it back but I know what that will lead to. I'll simply struggle against all odds and eventually have to give up. So I get up and walk towards a low wall a little way away from the others. I have left the children's swimming clothes there to dry and now I begin picking through them, stretching them out better in the sunshine. The tears are streaming hard and fast now that I am on safe ground.

Julius comes up to me, followed by Karin. When he sees my face he looks surprised but before he says anything I hug him to me and whisper that I am missing Gösta. Later on, we joke about how it is better that he and Karin only bicker and don't show any tenderness because then I won't miss Gösta so much. But Julius says that he had felt so positive after I had cried, as though I had cried on his behalf as well.

All the 'Ifs'

If only we had done this or that, then...

You can keep on like that. But there are no 'ifs' and there is no coincidence. There is no meaning. Life simply '*is*'. Cause and effect, ever since the first explosion, the Big Bang, up until today. Something happens because something happened before it that led up to it.

Everything has a story. If you follow events backwards in time you find that the lead up was logical and that there is an explanation for why everything happens. When two people meet 'coincidentally' there is an explanation for why they happened to be at the same place at the same time. One is not at a place out of pure coincidence.

If one wants to find meaning in life then one has to create it oneself because no one else can. If you do, then that may be good in your eyes. If you don't want to find meaning, then nothing will stop you from finding it meaningless.

One of the most meaningful things in my life was the prospect of having many children. I wanted my children to have siblings, to have each other and I wanted to follow their development into, ideally, contented adults. That trajectory of meaning has been amputated.

The question now is whether and when I shall create new meaning and why. The question of why is the easiest to answer: because of Julius and Karin.

The Country Cottage

I have mixed feelings as I follow the bends in the road. Auntie My has met Julius, Karin and myself at the station and now we are sitting in her rented convertible. I haven't been to the country cottage since Gösta's death and the last time I was there Gösta was with us.

It was in the summer two years ago. I was there with the children during our holiday back in Sweden from Singapore. We had had a good time – the children had been swimming and ran about as they wished in the village making friends with other children. I had my three children with me and My had her two. The cousins had explored the area together – an area that I assumed would become part of their future. They would come back here, just as My and I did, every summer since childhood. It was so special to watch the children play here as we used to. They went riding and climbed trees, stayed up far too late at night and slept far too long in the mornings.

Now we are returning, but with a gaping hole between us.

A little while later we are sitting in the village square with a beer each. Julius has got yet another ice-cream and if Karin had been beside us she would have got one as well. But she has scampered off somewhere. There are no cars in the village and the children are free to stroll about by themselves.

But after a while I ask Julius to go and look for her and he trudges off, dragging his feet. And shortly after he has gone, Karin turns up. So I ask her to go down the little alley in front of us and bring back her brother. Naturally, as soon as she has gone, Julius turns up. So I tell him that Karin is looking for him but that she'll no doubt be back soon. When she doesn't turn up, I get up and decide to go and search for her myself. I go to the far end of the narrow alley and reach one end of the village. No Karin. I call out her name as I walk but there is no answer.

I hold my breath and try to keep a grip on my thoughts. I will not allow myself to start fantasizing what could have happened to Karin. It is not until I have reached the far limit of the village that I hear her answer me. I start jogging in the direction of her voice, which is coming from the through road that leads the traffic away from the village. When I catch sight of her, she is walking towards me along the edge of the road and I run up to her and hoist her up into a tight hug. As I tighten my arms around her, I loosen my grip on my fantasizing.

I ask her why she left the village and why she went to the big road.

"Well, you said I should go straight on and look for Julius, so I did," she replies baffled.

'Oh kids!' I think to myself and squeeze her harder still. On the way back to the others, I tell her firmly that she must never go outside the village like that on her own.

My imagination has already gone wild; if I'd got there a couple of minutes later I would have called out but Karin wouldn't have heard me, she would have continued along the road and I would have continued walking around the village, she would have got tired and would have been lost and a car would have stopped beside her and ...

That evening in bed, I turn to look at Karin beside me and I pat her little naked body. It takes me a while to let go of the thoughts from earlier but I finally fall asleep with my fingers closed firmly around her warm hand.

When I wake up, I feel Karin's arm snaked around my neck. It is soft and smells sweet. I lie still and think for a while. Maybe I shouldn't because then I become aware that the fear from yesterday hasn't gone yet and that I am still tense. I turn onto my back and Karin pulls her arm away.

These 'ifs'. If I had come a few minutes later, if the wrong car had stopped, if, if, if ...

August Strindberg felt sorry for humanity. I know what he meant, or is it only myself I feel sorry for?

Today is a tough day. All these existential questions tumble over me and with them comes this agony. The constant worry about the children; it makes me feel terrible. And yet they are the reason I'm able to find meaning in life, a life that is so fragile but that I nevertheless take for granted somehow. I still have Julius and Karin but I am so afraid of something happening to them. Bizarrely, it is this that makes it seem as though Gösta is the safest of them all.

I pull myself out of bed, pull on a thin skirt and a singlet and go to the kitchen. It is quiet and empty. I'm the only one who is awake. I put on the kettle and pick up the packet of tea. By the time my tea and sandwich are ready it is already too hot to sit in the sun on the terrace, so I choose one of the chairs in the shade close to the house and relax in the morning breeze.

Everything is so familiar. There isn't a sound. Even the birds seem to still be asleep. If I listen hard I can just make out a whistle of the breeze, which is carrying the fragrances of dry grass and withered blossom. I can just make out our neighbour's roof which is almost completely concealed by greenery.

I so wish that I could fully indulge in all this magnificence. It is perfect – a quiet morning with a cup of good tea and Mediterranean sunshine. This place is a second home to me and has always meant relaxation. But instead, I feel uneasy because something is broken and I'm unable to give myself over to the sense of leisure.

Julius made the comment, "But it's good that you have me, Karin, Pappa and Kasper."

Later that day, I'm sitting at a thirty-five-year-old homemade kitchen bar looking at Julius who is sitting in an armchair a few meters from me. He looks straight at me and asks, "What's wrong Mamma?"
"Hmm? Do I look sad or something?"
"Yes."
"Gösta," says Karin, who is sitting on the floor doing a drawing.
She doesn't even need to look up to know.
"Yes, I say," and we need to say nothing more.

***

The worst thing of all is when everything seems perfect.
It could have been so perfect now.

***

Why is losing a child the worst possible thing?
Because it deprives one of the immense joy of being able to show the world to one's child and see him make it his own. It has nothing to do with all that nonsense about reproducing one's own genes.

***

It is a joy to see Julius meandering through shops stroking the clothing, testing colognes, looking around for more to explore. That is joyful.

***

159

Gösta is ever-present, on the train, by the table, in the car – but there is always an empty seat. I will not let go of Gösta. Never. He is with me and I will always make sure there is a space for him.

***

Having fun or a glass of wine with friends is time out from the rage, bitterness, agony and grief. Moments of joy numb me for a moment. Don't think that I can forget just because I laugh and mess about. Never.

***

I ask my sister My to change the ring signal on her cell phone from the one she had in Singapore. Every time I hear it, it is as though I'm about to hear news of Gösta.

Recollecting the Summer of 2003
(The Last Time I Remember
Thinking About How Happy I Was)

It is 3rd June 2003. The heat is tropical and the small lake is glittering in the sun. I am suddenly struck by the thought that I am happy. I stop for a moment and forget about all the schoolchildren and take note of how good my life is. In only a week, I shall have completed my year as class teacher and right now I am enjoying a day on the beach with my pupils. They are excited about the coming summer holidays.

In just a few days' time I shall attend the end of term celebrations at Julius' and Gösta's school, their last day at Grillby School. And then I shall pack up our home because we will be moving to Singapore, seeking new adventures. My mother and father have followed the whole build-up and are delighted for us. They are already hatching plans to come and visit us for Christmas and to then continue on a trip around Southeast Asia. My mother is so looking forward to it, she has told me, because when she was young she travelled in both Japan and Siam, as Thailand was known then.

Everything feels so good just now.

What, then, is happiness? I suppose it means that one is free of troubles and that nothing is weighing one down. One has control over the situation. But it is more than simply being in control. It is also knowing that nothing is painful and that whatever one chooses to think about, one can still feel glad. There is no need to steer one's thoughts; they can run free without colliding into something that must be avoided.

That was how I felt, full of anticipation and contentment.

I shall never again be able to feel that complete sense of happiness, allowing my thoughts to hover unfettered without risking that they will run into an abyss. I will maybe be able to find pleasure in small moments here and there and I shall just have to keep hold of those.

***

If I should live to be a hundred years old, I must never forget Gösta, the way he looks, sounds, smells. Let me always remember my nine-year-old exactly as he is!

***

Many people, even those who were not directly affected, have said that the tsunami made them stop and think about how they live their lives and relate to their children. Many say they have not appreciated enough. But I really did. I was so conscious of the fact that I had three wonderful children. I appreciated them and I felt so rich because of them. But I had my fears about what could happen to them, and I had fears about what could happen to Gösta.

***

I don't want to evaluate what came before and what has come after, but one outcome of Gösta's death is that death is no longer something distant or theoretical – not for me, and not for Julius or Karin either.

Life treats people so differently. There are those who have lived most of their lives without ever coming close to death; they have never lost a close relative or friend.

Maybe they only experience this well after middle age. I had never come close to death until I was adult.

Julius and Karin have experienced a death at such a young age. How will this affect them?

***

The photos of Gösta are becoming too painful. I want to take them down. I get so furious when I see pictures of him. They mock and ridicule me. It is as though I am famished and am only allowed to look at pictures of delicious food that I know could give me pleasure, strength and a will to live. The pictures of Gösta cannot satisfy my hunger – they simply whet my appetite. So I do not want to look at them.

Instead, I must look straight ahead and view life as it is right now. I must stay in the present and avoid looking sideways or backwards. Ahead of me are Julius and Karin, my home, Kasper, my job. Looking backwards is pointless nostalgia. Gösta is always with me anyway.

***

Each person has their own color. In the meeting between two people, their colors blend and create a new nuance that you either like or you don't.

***

It is early August and it is still hot. This summer has beaten all the records. I leave the verandah door open all day long and the natural world is starting to move indoors – the sitting room is full of small bugs, dragonflies and samples of vegetation.

One evening a small frog hops into the sitting room to Kasper's evident delight. I manage to get myself between dog and frog and capture the latter in my hands where he has pressed himself up against the bookshelves. I carry him carefully outside and put him down where there is still some moisture in the soil.

Julius and Karin are with Bertil and my time is my own. I want to sit down and write but it is too demanding. I would rather try to be here in the present moment. I put on my blinkers and repeat to myself, "here and now, here and now."

The photos of Gösta are still hanging on the walls. I haven't the heart to take them down. I just look down when I pass them so I don't meet his eyes. Looking at those photos is like the memory of having been in love. They provoke a longing for a time that has gone forever. I despise the nostalgia that I feel. It is absurd. I snort at the photos and mutter, "here and now, here and now." I want to enjoy the last of the summer and roam my garden barefoot and lightly dressed, watering my flowers.

Right now I have no strength for grief. The grief is there whether I make an effort or not. It has become cemented in me and is intertwining itself with my sense of self.

The very thing that I have dreaded is now becoming a reality – I am getting used to the fact that Gösta is gone. I knew it would happen and that this would ease the grief, but I didn't want it to. I still don't want it to. I want the grief to weigh as heavily as before.

Actually, I realize that in many senses it is just as burdensome and always will be. The difference is that I realize that I have to take a break from it sometimes. I have chosen to continue living but not as some kind of zombie. I have chosen to try and live in the present and this means I must push the grief behind me even though it may lurk over my shoulder, only a flash of thought away.

The Physiology of Thought
(It is Only Possible to Think One Thought at a Time)

The room of the present, the room of grief, the room of the past. There are almost certainly more rooms in our minds. Like the nurse who distracts the child from the needle by showing him a picture of a dog on the wall, I try to distract myself from thoughts of Gösta. But when I am alone, I can slip out of room of the present and drown myself in thoughts.

It is only possible to think one thought at a time. I can't think about making a cup of tea and think of Gösta at the same time. And one can only think as long as one is awake. This means we have some sixteen hours of thinking time per day. In other words, I have sixteen hours a day during which I could find myself in the room of grief. But many of those hours are occupied with the present, such as when the body requires attention: the need for food, drink, sleep, cold, warmth. When I am occupied with these things it is impossible to think of Gösta.

Nor is it possible when I am reading, listening, planning.

There are many reasons why I do not think about Gösta and there are ways in which I can ensure that I don't. I can decide not to. That sometimes works and sometimes doesn't. Sometimes my longing takes over and pushes all other thoughts aside.

***

If there is anyone that I knew I could always boast about my children to it was my mother. Now she too is gone, but perhaps it is best that way because I don't think she could have handled this. I don't think she could have lived with the knowledge that Gösta was dead.

***

Gösta is not real. He has not been a reality since his death. I have hated knowing that I would come to understand this. I have dreaded accepting that there could be a reality in which he played no part.

***

I fling Julius' trousers into his bedroom from the hall. Soon afterwards, Julius asks where his trousers are and I say he can look for them himself. I know he saw me throwing them into his room but he moodily begins searching, his irritation fills the whole room. A joiner arrives to do some work and he comments on the tension in the house. I pull him to one side and explain what has happened to us, that my son Gösta is dead, that he died in the tsunami and that this is why the atmosphere is so tense. The joiner's face contorts and glows red. Julius is standing beside me and I can feel the desperation swelling. But then Gösta is in front of me and I can feel the longing in me flowing through every inch of my body. He is so real, and I go forward to him and hug him hard until I gasp for breath and sob.

As I wake up I am aware of how I am wrenched from my dream but Gösta's presence continues. I am lying in bed, panting. My nose is blocked and my face is wet with tears. My limbs are heavy with adrenalin and I seem unable to move. I lie in this state for some time until Gösta gradually fades and I am left alone with my emptiness. Kasper has heard me and jumped up onto the bed to lie beside me. When I see him I want to throw him at the wall – he has disturbed me when I want to be alone.

Music

It is quiet in my home. There's no radio on and no CD player blaring. Sometimes I try the radio but I quickly tire of it and switch it off. I find all the advertisements and chit-chat between the music irritating. Occasionally, I'll put on a CD to listen to some carefully chosen song but since music tends to remind me of the past I usually avoid it.

I suppose I have never been a great fan of music. Like so many other children, when I was little I was persuaded to play an instrument, most of all I would have liked to play the violin, but first I had to play the flute and later the piano. But because I found it so hard to make my fingers stretch a whole octave both my teacher and I gave up after a term and a half. I have never missed it and I never came so far as to the violin. I have never identified myself with music or considered myself particularly musical. And so it is rare that I listen to music in my home.

It is quiet as usual on this Saturday morning as I sit at the kitchen table eating breakfast. The verandah door is open and I can hear my Indonesian wind chimes and bird chatter wafting in. That is enough sound for me.

I have an appointment with the hairdresser in an hour so I have plenty of time to read the newspaper and drink a whole pot of tea. Since this August day is warm and I have so much time, I decide to walk to the hairdressing salon, and although I dawdle I still manage to get there ten minutes early. So I figure I may as well go in and flick through a fashion magazine while I wait.

When I come in, all the hairdressers are bent over heads of wet hair or aluminium bonnets. I tell the receptionist who I am and she tells me that my hairdresser will be a few minutes late today. It suits me fine. On my way to the designer sofa in the waiting area I pick up a couple of magazines.

Everything is nicely under control. Here and now. I am holding myself to the surface; I'm comfortable in the sofa, the sun is warming the back of my neck, all the hair products around me smell good and I shall soon have a smart new hairdo.

I start flicking through the first magazine, which is for garden enthusiasts. Music from the 1980s is coming from loudspeakers on the wall and I realize that I do enjoy listening to good songs. I stop at the full-page spreads of gardens in full bloom and feel inspired and relaxed – until I hear Eric Clapton begin to sing 'Tears in Heaven'. As soon as he begins, I stiffen up and wonder if I should get out of here or at least put my hands over my ears. But actually, I want to listen.

Eric Clapton wrote this song in memory of his dead infant son. In his lyrics he declares that he does not belong to the heaven in which his son now exists. I can't help but feel that these words are also for me. We've listened to this song at home so many times over the years and have felt for this father, heard the pain in those words. I've always had to fight back tears when I've listened to it. This time, I make no attempt to fight them back.

I cannot resist the power of music. I cannot decide over it, it decides over me. It has nothing to do with whether or not I am musical or can play an instrument. It has everything to do with feelings and with tones that reach directly into the soul.

Now I understand why it is usually quiet in my home. I cannot deal with feelings over which I have no control. I must be able to choose when I'm prepared to allow them to capture me. I have been captured unexpectedly in this way by music before, when I heard Åsa Jinder singing and playing her key harp. So I know now that there is no point in trying to hold back. Gösta must take all the space and I must feel my grief. What else can I do?

***

The wardrobes must be cleared out. They are full of outgrown clothes and it's starting to get to me. I open the door of a wardrobe into which I've been chucking old clothes all summer long, and I pull out some bags, park myself on the floor and begin rummaging.

I soon have several piles in front of me. Winter clothes, baby clothes. Gösta's pile won't be very high because I've already got rid of most of his things. Then I pull out a little dress. It's a summer dress in yellow cotton printed with pink crowns. I pick it up and hang it in front of me. It has been washed but it is still stained. There is a large gray-brown mark by the waistline. It is the dress that saved Karin's life on Sunday 26 December 2004.

Bertil had run with Karin under his arm; he was exhausted but kept on running and the last thing he remembers before he felt that blow to the back of his legs was that he called out for help. Then he was bowled over into a somersault under water and he lost hold of Karin. He flung his arms around him and felt the cotton of the yellow and pink summer dress that she had been so determined to wear that day. He caught hold of the cloth and managed to get to his feet with his daughter in his grip. She was covered in sticky mud and she looked at him and asked:

"Pappa, are we going to die now?"

I repeat that four-year-old's question to myself, "Pappa, are we going to die now?" Although she was so young, she understood the meaning of these words. This child will know forever what death is and what it feels like to come close to it.

***

Kasper, what shall I do with you? You trail behind me wherever I go. You sleep on my bed, follow me into the bathroom, you lie beside the bath when I'm soaking in there, you greet me when I come home as though I've been gone for ages.

You are reckoned to be mankind's best friend because you are so affectionate and loyal. But me – I just want to be left in peace. I'm so sorry. It isn't your fault. It's mine. You are a such a lively, bouncy dog and you're so keen to please. But I have no space in my life for you. I'm so preoccupied with myself. I can barely cope with myself, let alone with you. You give me a guilty conscience because of how I feel about you, because I feel that you're in the way.

It's a vicious circle; you are in the way, you get under my skin and I push you aside and exclude you. Then I feel guilty because you don't mean any harm. But I feel irritated by you. Maybe I just project my frustration onto you. Maybe you should live somewhere else so that I can sort myself out.

***

Gösta is becoming less and less of a natural part of my life, our lives. In the early days of his death, he was at least with us throughout all those practicalities, like at his funeral. But he is no longer part of our everyday routine. It's only when we talk about his grave that he is with us again in some strange way. From now on we will only talk about Gösta on special occasions because he is no longer a part of our normal lives.

The day begins as usual with me sitting reading the newspaper and drinking tea. Then I hear the front door opening and Julius' voice calling hello. He comes into the kitchen and tells me he has a spare hour at school so thought he may as well come home. He sits down at the table to keep me company and we start chatting about nothing in particular. The clouds are hurrying past outside and when they cover the sun I feel a brisk breeze rush through the open verandah door. I shiver a little and get up to put on more tea water. Julius remains at the table. He still has more than half an hour before he has to be back at school.

I sit down again with a new cup of tea and we are both silent. I try to think of something to say but there seems to be only Gösta left to talk about. Gösta crops up whenever there is a gap in our lives. I think of how alone I feel when I long for him and how forced it feels to have to talk about him, but I manage to formulate a question.

"Do you think much about Gösta?"

It was sufficient. Julius looks at me with a sorrow in his eyes that has to be seen to be understood.

"Yes," he says. "Mostly I think about him when I'm at home alone. Not so much when I'm at school and there's so much going on."

Then he stops and the tears overflow. I am unable to be strong and consoling. I cry too.

Together, we try to find words for this sorrow but the phrases are halting and incomplete. I get up and fetch a roll of kitchen paper and hand some to Julius, who takes some to blow his nose on. Then we say nothing for a few minutes. At last I break the silence by ringing the school and explaining that Julius will not be returning to school today, and I see gratitude on Julius' face.

When I replace the receiver, I notice that the sun has started shining again, so brightly that it blinds me. It is making the oak leaves glimmer almost golden against the sky.

"Life would have been so good now. You and Gösta would have ..."

Julius doesn't complete my sentence. The words simply float out into nothingness and we fall silent again. Julius' cheeks slowly dry but his nose is still red. Then he says,

"Crying is like peeing. I mean the pressure builds up like when you need to go for a pee and then you feel relieved when it's over. Now I feel happy again."

But then more tears come and Julius looks almost surprised that he's crying even though he feels better. I say,

"I think it's relief. Sometimes we cry because we feel relieved. Our bodies and feelings have their own way of doing things and it's maybe best to just let them get on with it."

I've learnt that there is an art to listening to silence.

Belonging

I'm out meeting people.

"Hi! Wow, it was ages since I saw you. How are you? Great to see you, great ..."

Wine, cigarette, great!

A train home, lovely, home, peace and quiet, a buzz in my ears. Cold juice from the fridge, hang my smoky clothes outside, think about the evening. Get undressed. Wash. Walk down the hall towards the bedroom.

In the hall I catch Gösta's eye in a photo on the chest of drawers. Bang. 'Cinderella, wakes up from her dream.' The splendid chariot is transformed into a pumpkin, the shimmering dress, into rags. I flop into bed together with my sorrow. Gösta, how I long for you.

***

I'm sitting on the toilet. The toys in front of me on the bathroom floor have been lying there for days. Some of them are still wet from Karin's last bath. It's as though I'm seeing them for the first time. Many of them were Gösta's: Spiderman's motorbike, a little Action Man.

I realize that all these disposable toys have survived Gösta. All these plastic trinkets that are only supposed to last for a few months are still here, while my child is not. As they lie there, so nonchalantly strewn across the floor, they both mock me but also remind me of a time that did exist, a time that was so vital for me. I remember how much these cheap bits of plastic meant for Gösta and how they set off his imagination.

I get myself ready, step over the scattered toys and out into the present.

Katja

It is morning again and like so many mornings before, I am sitting alone by the kitchen table with a cup of tea and the morning newspaper. Julius and Karin are with Bertil this week and today might just as well have been yesterday or tomorrow. Time sometimes feels like a series of grids that can be laid one over the other. The difference between mornings has dissolved. The clock on the wall ticks monotonously and the fridge has stopped purring. There isn't a sound apart from the clock ticking.

I wonder if there will ever be a time when I step into another morning, backwards or forwards in time, or whether there is a place at which all mornings have been gathered together in a row. Parallel worlds. Could there be a time when one is no longer imprisoned by time?

This morning is similar to all the other mornings when I don't have the children at home and I expect nothing in particular of it.

But this morning will offer a surprise. On one of the pages of my newspaper I see a picture of a woman sitting side saddle on horseback. I recognize her instantly. It is Katja Schumann, a good friend from when I worked with the circus in New York, sixteen years ago. I say "wow" out loud and laugh. Katja is looking directly into the camera and laughing too. She is holding the reins with a firm grip and she looks radiant.

I notice the text beneath and I see that Katja is in Stockholm for a performance on horseback. I have to meet her! Memories of the Big Apple Circus tumble towards me like colorful baubles across the sawdust of the manege. I am transported over a bridge of thought back into the circus tent. I can smell popcorn and candyfloss, sawdust and horses. A thrilled audience has left an echo of applause in the empty benches and the lights still seem to beam in my eyes from the now darkened top of the tent.

My memories of that period of my life have been stored away in a compartment of my heart that I haven't opened for so long. But when I see Katja laughing with such warmth towards me, my heart opens and it all floods forth. It feels good.

I sit for a long time staring at Katja's picture, which is bright despite the fact that it is in black and white. The picture brings back the sound of trumpets, drums and whistles. I can't stop laughing. It is marvellous to be able to think back on such a life. How I have missed thinking about the circus; indeed, how I have missed the circus.

I get up, walk to the bookshelf and pull down my photo album from 1990. I snuggle myself into a corner of the sofa and begin turning the pages. It was so long since I looked at these pictures.

As I allow myself to be sucked back in time, I begin to understand how important it is to allow the past to exist and to dare to link myself with it and accept that it belongs to me. These are not the memories of another Ann; they are the memories of the person I am now.

Lincoln Centre, New York, August 1990. I lived there in a caravan that I shared with a woman who was working at the circus, taking care of the artists' children. I was there to visit my friend Marie-Pierre, whom I had got to know a long time previously when I was training in circus arts in Paris. She was a circus artiste in France and we were planning to perform together, but my life went in another direction.

I kept in touch with Marie-Pierre and when she was performing with the Big Apple Circus, I went to visit her. It was an extraordinary time. The photos show a potpourri of circus life, filled with colour and smiles.

There is a photo of me together with the clown "Grandma," my friend Barry. We are standing with our arms around each other, grinning at the camera. And there is one of him on a merry-go-round with his daughter. My memories of him come into focus. He lived at the circus with his family and he and I became good friends. When I left, he whispered goodbye and planted a light kiss on my lips. I wonder how he is now.

And what is Katja doing here. Has she separated from her circus director husband, Paul Binder, to concentrate on her own career? I feel curiosity welling inside me and decide to put all else aside and try to get hold of her. I swig down the remains of my cold tea, skip breakfast, get myself ready and then jump into the car and take off.

The show is to be held at Storängsbotten and at this time of day the traffic is flowing easily. When I arrive, I park the car by a fence and creep through a small opening in it. I see no one as I make my way towards the tent and inside it is dark and still. I am eager to find someone and feel a stab of disappointment at the silence. I head for the stables where some boys are dealing out hay and they look up at me as I approach. I ask if they know where Katja Schumann is and one of them tells me in English to follow him.

Outside is a large open area where a dozen or so caravans are standing in line. The boy who is leading me points to one in the middle.

"That´s where Katja lives," he says and then he turns and goes back to the stables.

"Thank you," I reply and walk towards the caravan.

It is unreal – that it should have been so easy to find her.

Less than an hour earlier I was as far from the circus chapter of my life as imaginable, and now I am standing right outside Katja Schumann's door. The door is ajar and

I can hear voices inside. I move towards the opening, knock cautiously on the door and wait. Inside,
Katja is talking to someone. She glances up in surprise.

I present myself and she looks at me for a moment and then her face lights up.

"Hello," she says, "come in!"

She turns to her visitor, who is already preparing to leave, and closes their conversation.

This is a precious reunion. Katja is just the same – full of energy and life. I ask and she responds. I have a thousand questions for her, and she has a thousand answers. We drink tea together and then she has to get ready for her evening performance. But before she goes, she gives me a long, hard hug and invites me to come to the circus in America again.

On my way back to the car, my steps are light like those of the circus princess I had always dreamt that I would become.

It isn't until I get home that I realize that I haven't told her about myself. When I was with her, I was thinking of Gösta and wondering how I could tell her, but I never took the chance. It was on the tip of my tongue, but I swallowed it instead. I don't know why I did that. Perhaps because I didn't want to ruin the moment. Maybe I just wanted to be happy for a while.

Sorting the Wash: Two Piles Instead of Three

Imagine that I used to feel irritated about having to sort the washing into three piles, and that I thought it was a nuisance that Julius' and Gösta's clothes looked so similar. Which underwear belonged to whom, and which socks were whose? And then there were all Karin's clothes which were at least easy to pick out but it took for ever to go through them all. I used to sort the clothes on the bed and there never seemed to be enough space for all of them. Everything would get mixed up and I'd get annoyed.

Now, when I only have to sort two piles. I so wish that I had to do three!

***

Sometimes I wonder about myself. I've got through yet another day of knowing that Gösta is dead. Gösta is dead. Gösta is dead. It's incredible that I can write these words, as though they were the truth. They are in fact the truth and I am nevertheless sitting here on the sofa on a Wednesday evening absorbed with what's happening in the election campaign. On Sunday I shall cast my vote for a government that is to hold power for four years. In four years' time Gösta will have been dead for five years and nine months. How will I feel by then? Will I be as astonished then as now that I have got through yet another day?

The Cross

Gösta is not dead. I've removed the cross that was hanging in the window and taken it with me to where I'm sitting on the bed. I weigh the cross in my hand; it is a large, black, iron cross with GÖSTA engraved across it. A cross – the ultimate symbol of resurrection. For me, it represents death. But when I gaze at the cross and read Gösta's name and try to repeat the phrase "Gösta is dead" to myself, a voice inside me protests. No, Gösta isn't dead. It doesn't feel as though he is dead. I know him so well and feel his presence so powerfully. He lives within me. I can hear his laughter in my ears, I feel the smell of him in my nostrils, I can feel his firm limbs in my hands, I can see his knobbed knees and round cheeks with my eyes. How can he possibly be dead when he is all around me, all the time?

I look more intently at the cross. It doesn't seem dangerous. I'm not cowed by the fact that it carries Gösta's name on it even though I know what that means. Because it isn't true. Gösta is more alive than many of those who exist materially on this earth. Gösta simply isn't here, and shit, how I hate that!

But he is not dead. I don't even know what is meant by dead. What is death if it isn't what I always believed it to be? I must look up the definition to find out what it actually means.

Is there such a thing as death? A plant grows, blooms, withers and dies. What is it that actually happens though? The moisture in the plant dries up but that is not death. It leaves a plant of a different color and it is dry and crumbles and falls to earth, spreads over the ground, sinks into it and is taken up by it. The plant is dematerialized but

180

it lives on in me because I can still see it, feel its scent and the rough stalk. So of course it is still living.

I find no answers in the dictionary.

A State of Being

Different dimensions. Giving birth to a child was like shifting into a new dimension, and losing one meant another shift.

I have two paths of thought. The first is more basic and perhaps more narrow. According to this way of thinking Gösta is dead and will never return. He is simply a memory. But according to the second way of thinking, Gösta lives on and is palpably present. I don't know if this way of thinking is more spiritual but it is greater, limitless.

There are powers that we humans possess that are beyond our comprehension. It is not a question of seeing Gösta again, because I realize I shall never do that. It is a question of how I experience him.

Human life is composed of experience and without this, we are dead. It is whether or not I experience Gösta that matters. Maybe this is simply a way to enable myself to survive and avoid being crushed by grief, but the important thing is that it comforts.

182

September 2006

Today it is one year and nine months since Gösta lived his last full day. I am bewildered by the fact that I am alive and that I managed to get through those days when we first got back to Singapore without Gösta. How was that possible?

How could I breathe, move, chew, swallow, knowing that Gösta was gone? How could I be so calm? Why didn't I go berserk, lash out, scream, kick – why didn't I go crazy? Will I later look back and wonder why I didn't go crazy when one year and nine months had passed? Maybe.

Maybe I stay sane because I move forwards and look downwards and keep watching my feet as they step onwards. I don't think of the fact that I will maybe have to keep walking forwards for another forty years like this. Maybe.

Sometimes life feels repetitive. It's like playing patience. You lay out the cards and each time, the pattern is different. Sometimes you're lucky and all the good cards come up and you have plenty of options. Sometimes the cards are not so good and there's not so much you can do. That's much how it feels when I wake up to a new day, "Ok, what will today be like?" But however it turns out, in many ways it will be just like all the other days. It just doesn't feel as though life has all that much to offer. The only thing I can be sure of is that each day that comes is yet another day since I was able to kiss Gösta.

Think of Julius and Karin.

I have always been materialistic. I don't mean that I've needed to acquire new things all the time but simply that I haven't been particularly spiritual. I used to only have this world as my reference point and I didn't need to think much about death. I figured it would come when it came. But now I've been forced to expand my horizons to include death as a part of life in a way that I never before dreamed of. And I don't like it.

I have always sought logical explanations for everything and have based my understanding of the world on the principles of natural science. When you die, it is finished, over, black. My vision was fixed firmly on the future

As science has expanded its reach and as I have come to experience the presence of my dead son, I have been forced to change perspective. Maybe there is something beyond what we can perceive. I didn't used to care what that might be and I was content with the world I could experience with my five senses. But after 26 December 2004 this no longer held and I was forced to search for Gösta in another dimension.

With one foot here in our world and one in the realm we call death, I have to raise my line of vision. Maybe it's because I can't bear to think of Gösta as dead that I must keep looking for him. I know he doesn't exist in our dimension, so it is only if I confine myself to these that he remains dead and gone.

But if I raise my eyes and try to see beyond, maybe I will discover something. How should I do this? Just as many people say they have found God when someone dies, I too can begin to discern something beyond our material world. But I don't want to.

I don't want to achieve greater insight or a deeper understanding. I don't want to be wiser. I don't want anything positive to come of this. I want to stay put in my comfortable material world in which everything was trivial and tangible. And I had Gösta.

Anxiety

Many of those who experienced the tsunami but didn't lose anyone now feel anxious about the wave. I understand this. I would also feel like that if I hadn't lost someone. I would have thought about what could have happened and would have had nightmares. But the worst possible thing did happen.

Bertil remembers that he saw the wave coming. He remembers houses and people being washed away and he heard the roar. I have no memory of it and can only imagine what it looked like. It is peculiar that I don't remember because I turned around to tell Bertil to hurry up and I ought to have seen it then. Is it possible that I completely blocked this out? Is this maybe why the wave itself doesn't worry me?

Italy

Half an hour before I should leave the phone rings. A stranger's voice says that he is interested in buying my cell phone, which I have advertised on internet. Half an hour later, I leave with a little more travel money than I expected.

It is late September and we are about to embark on the trip we planned back in May. I've just met up with my friend Lotta, with whom I was traveling, at the airport and we are going through the security check. We still have an hour before the plane takes off for Rome and Lotta needs to change some money, so we head towards the currency exchange counter. Then I hear a woman's voice calling out,

"I thought I recognized your voice! Hello Ann!"

I turn towards the voice and see a face I know so well but which also feels distant. I realize that it is Anki, from the Swedish embassy in Singapore, but can't quite believe what I see. It is as though a figure from some favourite book just popped up for real.

I am so glad to see her even though I can't quite make the connection – that it really is Anki. She gives me a big hug and I hug her too before collecting myself and introducing her to Lotta. It is when I say Anki's name that I confirm for myself that it is her. Now we are all three heading towards the currency exchange counter and Anki is explaining that she has just been to a meeting in Stockholm and is on her way home to Singapore. I mouth for myself, "On the way home to Singapore." When I think of Anki sitting on a plane headed for Singapore it stirs feelings in me, mainly positive feelings, which surprises me.

I ask her about various people we knew but she tells me that many of the Swedes we knew there have already left. It is a relief to hear that not only I but also Singapore has changed. I enjoy listening to her talk about Singapore, as if I had come to doubt whether I had actually lived there once. She makes me realize that I have lived there and that I have a place to stay in her home next time I go there.

It is this that makes me realize that it isn't dangerous to embrace the life I had there, and that I have a right to the time I spent there. Anki is a bridge to something I hardly believed belonged to me anymore.

When we are sitting on the plane later on, I think to myself how good it felt to meet Anki. It may be a long time until I see her again, but she is part of me and I suspect I'm a part of her too. I look forward to seeing her in Singapore some time.

The Fall

The fall has mercilessly arrived. Half rotted leaves have formed a wet brown carpet along the sidewalk on the street leading home. Although it's only four in the afternoon it is almost completely dark already. I'm walking the few hundred meters from the bus stop home after work. My thin shoes are making a sloshing sound and there seems to be no point in trying to avoid the puddles any more. It isn't raining any more but the air is damp and heavy all around me.

I'm looking forward to getting home and getting out of the raw autumn and into the warmth indoors. I shall put on the kettle, make a cup of coffee, light a candle and then curl myself up on the sofa.

I pull my jacket tighter around me and feel a rush of pleasure at the thought of the contrast between the cold outside and the cosiness indoors. I think of the taste of gingerbread and remember how, so long ago, I used to think the fall was magical with its new palette of colors.

The fall ushers people into their houses. I usually mark its arrival by stocking up on gingerbread cookies because they somehow signify the fall. We sit by the kitchen table drinking milky coffee and stuffing ourselves with gingerbread cookies. Candlelight makes it even more homely. And now, on my way home, that feeling sends a warm feeling through my veins, just for a moment. But only for a moment, because the feeling is also foreign and not really my own. It's just taunting me.

I'm not sure if I will buy any gingerbread cookies this year. It would feel a bit as though I was trying to kid myself that this fall is just like all the others; I'm torn between buying some and not buying any.

Bertil

Bertil and I have accompanied Julius to the psychologist. He has met Håkan a couple of times now, but sometimes I wonder if it is Bertil and I rather than Julius who want these meetings to take place. Julius doesn't seem all that interested and may only be doing this because we think it's a good idea. He is incredibly loyal.

It is empty in the waiting room where we sit waiting for Julius to come out, and the rain is spattering on the window.

"You're welcome to come home for dinner this evening if you don't have anything else planned," says Bertil as he continues to watch the blustery dusk outside.

"Sure," I respond slowly. "I only planned to go home and take a bath so that would be nice."

"Good. We can go straight home to my place and make food together."

Soon Julius joins us in the waiting room and Håkan is right behind him. They ask us to come into Håkan's room and so we all go in and have a fruitful talk together.

The atmosphere in the car on the way to Bertil's house is easy going. Julius knows that he is welcome back to Håkan at any time. That is enough for him and he knows he is free to choose as he pleases.

While the lasagne is bubbling in the oven, I sit at the table with a glass of wine and listen to the soft music. The children are busy and Bertil and I are chatting. It's a pleasant Friday evening. The music stops and Bertil goes to find another CD. He puts it on and I recognize it so well. A curtain goes up and I find myself sitting on the sofa in Singapore. In front of us stand Julius and Gösta, ready to perform the dance they've been practising to this particular George Michael song. As the music starts, their arms start flailing and their bodies spin as they laugh and fall about.

"No!" screams Gösta. "We did it all wrong. We'll have to start again!"

"Ok," says Julius.

And they start all over again.

When they have finished they bow low several times and we clap wildly and do wolf-whistles. The boys then collapse into sweaty heaps on the sofa and as I begin smothering them with kisses, Karin leaps up and takes the floor to do her own dance as George Michael sings on.

"What's happening Ann?" asks Bertil when he sees me with elbows rooted to the table and gaze fixed on the dark window.

"Don't you remember this music?" I say, and I remind him of the boys' dance. "It's so damned painful listening to this."

Bertil sits down opposite me and says nothing for a time. Then he says,

"Gösta can really catch you when you're least prepared. And the worst thing of all was greeting his coffin at Ärna."

I shake my head. I can find no words, but words don't seem necessary either.

We remain silent until the egg-timer tells us that the food is ready and Julius and Karin come expectantly into the kitchen. They are hungry and Bertil is on his feet already getting dinner ready for them.

Visiting the Grave with Karin

It's a long time since we last visited Gösta's grave. It's already late October and the summer flowers we planted must have all shrivelled up by now. Karin and I decide to go and tidy everything up and make it beautiful for the fall.

It is Sunday, the sun is out and the woodlands are a mosaic of different colors. Karin and I drive to the florist and there we busy ourselves checking out all the different plants and flowers. I have to keep reminding myself why we are here.

An hour later, we are on our knees digging around in the soil over the grave. Karin suddenly gets up and says,

"I haven't thought about Gösta for ages!"

I see guilt on her face. But her voice betrays more surprise than guilt – as though she only just realized that she hasn't thought about her dead brother for a long time.

"But Karin, that must be nice? It must mean that you've been thinking about fun things and that you've been feeling happy."

"Yes, but I haven't thought about Gösta at all!" she insists.

"That doesn't matter. You don't have to think about Gösta all the time. It's okay to think about other things too. You'll never forget Gösta even if you don't think about him so much."

Karin crouches down again and resumes her digging. She says nothing for a while but then she says,

"But now I'm thinking about him masses!"

An Ordinary Saturday Evening

It could just as well have been a Friday evening or even a Tuesday evening. It is half past six, Karin is watching the children's programs and Julius is playing a computer game. I'm in the kitchen making dinner. I am crying. The tears stop for a while and then they start again. Nobody sees and nobody knows. I lean my forehead against the fan hood and let the steam from the saucepan rise into my face.

It is when I'm like this, away from the world, that I can give my feelings space and can let go of my inhibitions. I feel how my heart contracts and how the pressure builds up around it, reaching up into my throat before it takes over and I have to let the sobbing out. The sound is drowned out by the spitting of the onion frying in the pan in front of me.

I wonder what I'm actually crying about. Is it because no one has called to ask what I'm doing this Saturday evening? Is it because Julius has to play his computer games alone? Is it because I'm now alone or because Gösta has died?

Whatever the reason, I feel delicate and tired. I find it irksome to have to bear myself up and feel all the restrictions of my cell. I want to escape.

The tears stop and I begin imagining how I take off. I travel to Africa and set up camp out in the bush somewhere. It is hot and the smells are unfamiliar: dry earth, exotic vegetation, goat droppings. I see a huge sun sinking low behind a distant mountain and fires begin to dot the landscape with light. I have left behind all the demands, endeavours and disappointments of my past and am free to live in the present. No past, no future.

Discussions with the Criminal Investigation Department

He is sitting at the other end of the telephone line looking at a picture of Gösta in front of him. The policeman informs me that the body is intact, all body parts are together, that there is some hair still on the head and that there are no visible signs of injury. Nor are there any visible features; the body is severely decayed. The policeman strongly advises against looking at the photos. He has photos of Gösta's body lying in front of him, at the other end of the telephone line.

***

I am talking to my friend Janne on the phone. We are discussing our children and how they are getting on with their activities, friends, schools. I tell him that Julius has just started at an English medium school where half the lessons are taught in English. I tell him how pleased I am that Julius has this chance to develop his language proficiency. I tell him that the school has an international feel to it and that it reminds me of the school Julius attended in Singapore. "The school Julius attended in Singapore."

But what about Gösta? Shit. Has so much time already passed that I only mention Julius and forget to say that Gösta attended that school as well?

PART 4
2007

The Train Ride to Gothenburg

It's Friday morning and I'm on the underground headed for Stockholm central station. The weather is miserable but it not as cold as it usually is in early January. I've got plenty of time before my train leaves for Gothenburg. It's nine o'clock in the morning and I have time to nose around in the shops for a while before heading over towards my platform. The train is already standing there so I dig out my ticket from my bag and check the number of my seat and wagon. I hope it will be a window seat on the left side so that I can see our old house when we pass Grillby.

I'm in luck! My seat is right against the window and I'll get a full view. I heave my bag up onto the overhead rack and nestle into my seat. I've been looking forward to going to Gothenburg for Sofia's fiftieth birthday party. And I love travelling by train, looking out at the landscape and feeling the distance between the places on is travelling from and to.

As the train pulls out and slowly begins to gather speed I look back at the city. Despite the lid of ice that has settled on the canal there are still some boats in the water. We are picking up more speed and passing Tomteboda, Solna, Sundbyberg. Stockholm is not so extensive and we are soon leaving the city.

The landscape opens up and forests and lakes flash by. I know this route so well. When we pass Ekolsunds Bridge I sit myself up to get a better view. We are approaching

Grillby and I need to be ready. The train will pass the oak forest and gravel track that takes you all the way up to our old house. It's quite a hike but Bertil and I strolled along it together many times. The gravel track runs parallel to the train track.

The landscape is now completely open and I see Ekström's fields passing. Why do we have to go so fast? I don't get time to look properly. Up ahead I can see the Ekström's farm and the track that leads up to our house. And there, there is our house! The lawn in front of it is still green. I see Julius and Gösta jumping up on the back of the lawnmower while Bertil drives. The boys are beaming. I see the chicken coop where all our fine hens used to strut about pecking at the ground. And there is the little stable where we kept the goats. And the path to the toboggan slope (which doesn't look nearly as steep or long as I remember). I can see us now, after a lazy Christmas Eve breakfast in front of the open fire trudging off to the toboggan slope with the sun glinting on the snow. The last thing I see is the impressive chunk of rock that sticks up in the middle of the field, where we used to walk with the goats bucking around us like dogs.

It's like looking into one of those old-fashioned tittägg, peeking into a whole other world. I thought it was only going to be fun looking out at our old haunts, but the memories cave in on me and I tumble backwards over a cliff-edge.

The train drones onwards and I get up and make my way along it so as to get one last glimpse of my old life, that I once said good-bye to without ever realizing how definitive that farewell would be. It disappears behind the train.

I remain standing with my forehead glued to the glass of the window wondering how we could ever have left such an idyllic way of life.

The longing for Gösta
I long for Gösta
The longing for death
The longing to death
I long for my time to die

July

Separation anxiety. I don't want to leave my old life. But I know that I would have to and that I would eventually have to accept a life without Gösta and even learn to like it if I was to survive. I had to fill that life with something, fill up the eternal vacuum that Gösta had left. But I didn't want to.

I didn't want to live a life without Gösta and didn't want to accept that it was possible to live without him. I didn't want to let go of my life as the mother of three. I didn't want to move forward.

Move forward. What does that mean? Of course you move forwards because you can't stop time. You move forwards whether you want to or not. How could you not do so? For every second that passes, I move forwards and in so doing move one second further from Gösta. It has nothing to do with what I want or don't want to do. There is no choice. The only choice I have is about *how* I move forward.

But what should I do if I see myself as a light-hearted person who is full of life and visions for the future, as someone who likes to enjoy herself and spend time with friends but who is now forced into this? It's not easy to continue as before, but nor is it easy to keep grieving the whole time. I feel like an old woman – like I thought I'd be when I was eighty-seven and sat waiting for the grandchildren to call.

Here I am on a Sunday morning in mid-July and am not even hungover. I'm leaning back in my red wicker armchair in my sunroom enjoying a cup of tea. I've put on some meditative Indian music indoors and I can smell coriander and thyme from the garden. I have a shawl over my shoulders and sheepskin slippers on my feet. Beside me lays my book and my glasses. It is perfectly still.

I look out at the garden and watch the branches sway. Nature is in full swing out there, and the sun is out one minute and gone the next. Today I'm free to sit here all day if I want. I tend to feel like an old woman when I let my thoughts roam freely because then they tend to flow backwards rather than forwards. I think more of my life as it *was* than my life as it will be.

But maybe it's okay to sit as I'm doing? Maybe it's nonsense to expect life to go at full speed all the time and that everything should be such fun. Maybe this is also a part of life.

I can never escape my life – a life that feels like a sled that I need so much energy to pull behind me. It is a sled full of memories, wonderful memories, but they weigh so much. The sled moves forward but only on certain conditions; it needs a downwards slope, a tail wind and slippery ground. Then I can forget for a while, feel the wind beat my face and be fully in the present. If conditions aren't ideal then I don't have the strength to pull my load. What will happen if the sled quite simply stops moving? If I let it stop?

I think of the concept of "getting over" sorrow. We get over some things within a few days – things like missing the bus to see a movie or finding we've been short-changed. Other things require more time, like getting fired from a job, being burgled or spilling red wine on a favourite dress.

One thing that takes years to get over is betrayal – betrayal by a friend whom you've entrusted with a precious secret, or betrayal by a lover. It can feel impossible to get over something like this or it may take years. There are those who commit suicide in the wake of a betrayal.

However, some people say, perhaps twenty years later, 'I got over the sense of betrayal. It took a long time and I never believed I would, but I have. Now I have recovered the sense of friendship and love.'

In other words, whatever happened twenty years ago doesn't hurt any more. It was painful then, but not now. It's possible to think about it without feeling pain and it doesn't matter anymore.

Is it possible, in a similar way, ever to get over the pain of losing a child?

Will I ever be able to say that it no longer feels painful that Gösta is dead? Will I ever be able to say that it no longer matters that Gösta died?

Is this a question of accepting destiny? Finding oneself? Reconciling oneself?

I've always previously felt able to get myself out of situations that weren't to my taste, but now I can't. The only way out will be in death. Death will release me.

Garden Party

The day has finally arrived. It has been warm weather for almost a week now and the sun is still glowing in the clear, blue sky. Today I shall hold a garden party. Everything has been arranged and I am sitting on the verandah with a cup of tea. It is still early – the clock hasn't yet struck nine. The trees are throwing shadows over the lawn and the flowers are still in bloom. Only the lilac has faded. Behind me, the hammock is swaying in the morning breeze.

I pick up my phone and ring Julius on his mobile. He answers after one ring signal.

"Hi Julius, I can hear that you're on the bus."

"Yep."

"Did it all go okay? What a noise."

"Yeah, there are so many people on the bus."

"Will they pick you up when you get there?"

"Yes, they'll pick me up from the pier."

"Great. I hope you have a super weekend out on the island and you must say hello from me."

"Thanks, I will. Hope you have a good party too. When does it start?"

"Sometime this afternoon. Your auntie My and Marcel are coming to help out around lunch time."

"Sounds good."

"Okay Julius, bye for now. See you tomorrow. Hugs and kisses."

"Bye. Hugs."

I switch off my mobile and look out over the garden with a smile.

"My little Julius."

My little Julius, I repeat to myself. I can see him sitting there on the bus, full of anticipation and I feel pleased for him. I hope he will enjoy himself with his friend Martin. I sit for a few minutes and feel my eyes beginning to sting. From the pleasure of hearing Julius, I find myself flung into grief. Gösta. Why can't Gösta experience this summer? Why isn't he also sitting on a sweaty bus on his way to visit a friend?

OK, let it come. My grief swells inside me, and I understand how close pleasure and pain lie. A minute before, everything was fine. I had my head above water and was pleased. But everything is so fragile. It was precisely that feeling of contentment that made it so easy for everything to crack. It was the knowledge that Julius was doing well that brought it all crashing down. Agony and ecstasy, hand in hand. I can't distinguish the one from the other.

I remain sitting, weathering the storm for a while before I get up and go to the bedroom. On the way, I pass Gösta's photograph and the attack intensifies. Once I'm inside my bedroom, I open a drawer to take out a jumper and try to think of which colour I want. The cerise or the ice-blue? I choose the cerise and feel grateful for the fact that there are still trivial matters that can engage me. I put on the jumper as I am walking to the bathroom, where I blow my nose and begin ascending to the surface again.

Visiting Astrid Lindgren's
World and Gösta's Grave

The alarm clock shrieks. It's eight o'clock and Karin, Julius and I stumble out of bed and get ourselves cups of tea. Karin is too sleepy to eat anything so I take some fruit from the fridge and put it in a plastic tub for later. Soon, we've poured ourselves into the car and are on our way to Astrid Lindgren's World in Vimmerby. We have decided to stop by Gösta's grave on our way.

When we're on the highway heading towards Grillby, the cloud cover starts to break up and since it's still so early in the day there is hardly any traffic. The sun follows us all the way to the Grillby exit and along the road that leads to the church. We reach the crossroads and I turn onto the gravel road and slow down more than I really need to. There are paddocks with grazing horses on each side of the road and ahead of us are the huge trees that shelter a farmhouse below.

I notice how beautiful everything seems here. We roll slowly towards the farmhouse and the pens full of sheep with their lambs. We slow down even more until the car is creeping along and we can see the church and its stone wall through the trees. This was the way I used to drive home. I used to live in these picturesque surroundings and we used to walk down to these sheep pens and let the children feed the new lambs.

We continue along beside the sheep until we reach the parking area in front of the church. Karin has been asleep almost the whole way but now she wakes up and asks eagerly whether we are at Astrid Lindgren's World already. But when she looks out of the window, she remembers that we planned to stop en route and visit the grave. I switch off the engine and we get out.

All of a sudden the situation becomes absurd. Here I am with my children on this perfect morning in the middle of their summer vacation. We're on an outing together and the children are bubbling with excitement. Everything is perfect. We're just going to make a quick stop on the way. We are just going to stop at the grave of my other child.

My two children are walking beside up to the grave where my third child rests. When we arrive at the graveside I stand in front of it and look around me at the area. The scenery is exquisite. Now I have all three of my children close by me, but only two of them can enjoy the summer weather and the fact that they are on vacation.

Always this contradiction; grief bursts the bubble of joy. Grief and joy, conjoined like Siamese twins. Wherever there is joy, there is also grief. But the reverse is not true. Joy lets you down, but when grief rolls in, it does so alone.

Today is the first day that I have been truly sad when I've stood by the grave. It's as though this is the first time I truly appreciate what it is that I am standing beside. I look at Julius and Karin standing silently as they look down on their brother's grave. 'Our beloved Gösta Schmidt, born 20 February 1995, died 26 December 2004.' Spiderman is clambering up the gravestone in his red and blue outfit.

The contrast is so stark. My three children are here, and we're out on an adventure, but Gösta can't accompany us. The sun is winking through the foliage that shades Gösta's grave and where the rays touch the stone it glitters.

I rest my arms on the shoulders of Julius and Karin and begin to lead them back towards the car. The chasm between the two of them seems bottomless, and terrible.

(I don't believe)
I believe in God
I don't believe in God

I don't know
I know of God
I know nothing of God
I hope for a life after this one

***

After a long day at Astrid Lindgren's World it feels so good to put my feet up in the tiny cabin we've rented for the night. Julius and Karin think the cabin is far too small. They had envisaged something more luxurious but I think it's just fine. It is a bit cramped so it'll be interesting to see what happens when we pull out the sofabed because the floor will completely disappear. But first we must go and find something to eat at a nearby restaurant. The children have decided that they want to eat Chinese food.

When we step out into the narrow alley beside our cabin it is raining hard. This is the alley which Kalle Blomkvinst, one of Astrid Lindgren's characters, had run around in. The houses we pass are so quaint. It is easy to see how this place inspired Astrid Lindgren.

It is raining and thankfully we came equipped with rain clothes and umbrellas because when we reach the square we see posters advertising the evening's outdoor performances. A rock band is already warming up, but we head for the restaurant first.

We order all kinds of dishes to share and we pick them over with our chopsticks, laughing at each other and joking as we eat. It warms my heart to see Julius and Karin enjoying themselves and each other and it's good to know that we can have a good time together. I was worried that Julius would think it was tedious to come with Karin and me, but I can see that he appreciates the feeling of being together like this.

"You are the world's best mum," says Karin determinedly.

"Yeah, you are," Julius agrees.

Life is worthwhile after all.

When we leave the restaurant the rain has already stopped, though the air is still damp and heavy. But we walk over to the square where three brave souls are doing their best to make the otherwise empty rows of plastic seats look full. The band is playing with gusto and we listen to a couple of songs before joining the other three in the audience in an enthusiastic applause and then making our way back to our cabin.

***

That night, I meet Gösta.

I am running down a slope towards a field together with other relatives or the dead. In the field are all the coffins. They are spread out but they belong together. I can't see how many there are but it looks like more than ten anyhow. The field is brilliant green and full of flowers, and all around it is birch forest.

Halfway down, I notice Gösta. He has grown so much. He's leaning against a lamppost and he nods towards his coffin down in the field. I run over to him and say that I haven't seen him since the tsunami.

"Where have you been?" I wonder.

He tells me he has been here all the time but I just haven't noticed him.

In the morning I wake up with the feeling that I have all three children around me.

Looking Back: Gösta's Many Lives

Gösta was going to visit Granny and Grandpa:

The sunlight was making the snow outside glisten like diamonds. Granny and Grandpa had arrived the day before and had made themselves at home in the cabin in our garden. I was sitting in the kitchen with my tea while Julius and Gösta tore about in whirlwind of excitement.

"Can we go in to Granny and Grandpa yet?" Asked Julius, yet again.

"Okay then. They're probably awake by now," I responded as I looked at the clock and saw that it was after all nine thirty. It wouldn't matter if they got woken up by their grandchildren. What could be better than being woken up by your grandchildren the day before Christmas Eve?

"Yippee," screeched Julius and little Gösta laughed too, following his big brother's lead as always.

"But put jackets on. It's freezing out there!" I called out to them as they raced each other to the front door.

I walked over to the window and looked at the outdoor thermometer. It registered twenty-two degrees below zero.

"The door is locked," Julius called out. "I can't open it."

"I'll help you," I said. It was far too stiff for a five-year-old to manage.

As soon as the door was open, Julius bolted without a thought for Gösta, who was still trying to get his shoes on.

"Come on Gösta, I'll help you with those shoes," I said, taking his small warm hand in mine. "Ok, now you can go and see Granny and Grandpa too," I said as I opened the door for him and closed it behind him. It wasn't far to the little cabin.

A moment to myself. Bliss. I put the newspaper under one arm, grabbed my teacup and headed for the sitting room. The whole room smelled of pine from the Christmas tree and the fierce cold outside had put a chill on the air even inside. The morning sun was bathing the floor in pools of orange light. Outside, the snow-covered fields stretched all the way down to the railway line and I could see a train, like a blue ribbon, speeding by on its way to Gothenburg.

I stood there looking out of the window for some time before sitting down with my paper. It was so peaceful. This was my home.

By the time I'd finished the newspaper, there was noise at the door.

"Hello, is anyone awake in here?"

"Good morning! Yup, we're up. Bertil is in the shower and I'm in here with the newspaper and my tea."

"Goodness, look at the light in here," said my mother as she came into the sitting room.

"Mm, isn't it gorgeous? Perfect for Christmas. How are things in your cabin? Are Julius and Gösta still in there?"

"Julius and Gösta? Only Julius. He's in there with Grandpa, having a coffee!"

"And where's Gösta?"

"I don't know. It was only Julius who came in."

"What? Then where is Gösta?" I said in alarm, and hurried out into the hall.

I didn't have time to hear what my mother said to me before I had my boots on and was out. I pulled my dressing gown tight around me when I felt the cold bite at me. Then I took off at a jog up in the direction of the gravel road. I kept looking all around me as I ran past the few houses that constituted our hamlet. I couldn't see Gösta anywhere. I passed the letter boxes at the end of our land and continued along the road that wound its way up to the church and there, way up ahead near the church

I could see a black dot inching its way along the road.

When I reached him, Gösta turned around and looked at me in surprise.

"Little Gösta, where on earth were you going?" I asked him.

"I'm going to Granny and Grandpa," he said brightly.

"Oh but sweetie," I said as I picked him up and turned to go home. "I'll show you where they are."

***

When Gösta got something stuck in his throat:

It was a midweek evening. Julius and Gösta had gone to bed and Bertil and I were sitting on the sofa chatting when we heard footsteps on the stairs and then saw Julius at the door.

"Gösta sounds strange," he said.

We both leaped up from the sofa and rushed up to the boys' room. Gösta was lying in his bed sobbing with mucus running from his lips.

"He's got something stuck in his throat," screamed Bertil.

I grabbed Gösta, flung him onto his stomach over my knee and thumped furiously on his back. A three-centimeter-long tin soldier projected from his mouth and Bertil and I looked at one another and then at Julius. That evening we filled a large sack with small toys and dumped the whole thing in the garbage.

***

When Gösta walked along the road:

It was early spring and I was down on all fours by the flower bed clearing out the previous year's dead leaves. Julius and Gösta were out playing with the boys next door.

Gösta had just learnt how to walk and he was toddling along uncertainly behind his big brother. When I stood up to stretch my legs, I saw Julius and Linus walking up the driveway.

"Where is Gösta?" I called.

"Don't know," they replied in chorus.

"But he was with you just now."

The boys looked at me without appearing to understand. I looked about but couldn't see Gösta anywhere.

"Where have you been?" I wondered. I was beginning to feel worried.

"We were down there playing by a puddle," said Julius and pointed down the road.

"What? Have you been by the road? Cars drive along there."

"But we weren't on the road, we were beside it."

"Was Gösta there too?"

"Yes."

I turned and began hotfooting it down towards the road. And there he was. In the middle of the road I saw Gösta wandering along with a stick in his hand that he was dragging along the asphalt behind him. I rushed towards him, swung him up into my arms and began striding back up towards the house. While I was walking I shuddered when I heard a car speed past along the road below.

***

When Gösta got stuck in the apple tree:

It was Easter time and we were all sitting eating a special meal of lamb together. My parents had joined us at Bertil's parents' house up in Falun. Auntie Eva, Bertil's sister, had brought her husband and children along too. The children had all sat obediently through the meal but as soon as they had finished eating they began begging to be allowed to leave the table, so we told them that was fine.

After a few minutes Julius came running back to tell us that Gösta was hanging in a tree. My father was the quickest to set off, with Bertil and myself close on his heels. When we reached the tree we saw Gösta hanging high up in it from one foot, which was caught in a fork in the tree's branches. He was upside down and his arms were waving wildly. The children explained that he and his cousin Johanna had been sitting on a branch when Gösta suddenly lost his balance and would have fallen if his rubber boot hadn't fastened in the fork from which he was now hanging. It was my father who reached him and brought him safely to earth.

***

When Gösta fell down the basement steps:

We were sitting in the garden and hadn't yet got changed into our finery for our friends' wedding when Julius came rushing towards us, his face flushed and his voice wild as he screamed,

"Come, hurry up, Gösta has fallen down the basement steps."

We were up and running after Julius in no time. We rounded the corner of the youth hostel that we were staying at and reached the back side of the house where a stairway led down into the basement. When we reached the entrance we saw Gösta lying at the foot of the stairs. He was drowsy. I carried him up to our room and lay him on our bed. His face and hands were grazed. Bertil sat down on the bed beside him and tried to get him to respond. Gösta looked up at us but he neither cried nor spoke.

Bertil took out his cell phone and rang the hospital where they advised us not to let Gösta sleep too deeply, but that he should lie down and rest. We asked Julius what had happened and he explained that he had been standing at the bottom of the stairs and Gösta had been crouching down at the top beside the stairs. He had leaned forward to see Julius better and had lost his balance and tumbled down around one and a half meters head first straight onto the concrete floor below.

Gösta lay on the bed and dozed. He seemed a bit dizzy but we managed to keep contact with him and after a couple of hours he perked up and we still had time to get dressed up and go to the wedding.

***

When Gösta cycled to Oscar:

Gösta was standing by the kitchen table battling with a belligerent sheet of wrapping paper. He was determined that he would wrap up his present to Oscar by himself, ready to take to the birthday party. When he finally conquered the paper, he put the package into a bag, went out to his bicycle and fastened the bag onto the back,

Gösta was going to cycle over to Oscar's and while he was at the party, Bertil, Julius and I would go shopping. When we reached the church we saw a little being pelting along on his bike up ahead. We drove up behind him and asked where he thought he was going. He said he was on his way to Oscar.

"But Gösta, he lives over there," I said, pointing to Oscar's house some hundred meters from ours.

"No, I'm going to the Oscar from my daycare, and he doesn't live there."

"No, you aren't going there Gösta. The party is at our neighbour Oscar's. Come on, jump in and we'll drive you over there instead."

Gösta hopped into the car and Bertil flung the bike into the back of the car. A few minutes later we passed the church and swung out into the traffic on the road towards Enköping.

"Holy cow," said Bertil. "It was lucky we left when we did and caught up with Gösta. Otherwise he would have been out on this road and heading towards the wrong Oscar – and he lives about ten kilometers from here and he doesn't even know where he lives, he would been so lost out here. Imagine."

"I know," I replied as I watched the cars flying by.

"Sometimes luck is with us."

***

Gösta at the boat fair:

One of the first signs of spring is the Boat Fair. There was still snow on the ground but the sun was warm and the sidewalks were becoming visible again. Julius and Gösta were itching to get out and run around after sitting in the car for more than an hour from Grillby to Älvsjö. They shot off and found a heap of granular, muddy spring snow to play on. We were all dirty and wet when we entered the exhibition hall and then we had our hands full keeping track of one five-year old and one three-year old who thought nothing could be more exciting than running up all the ramps to the boats. We managed through about half the hall but then Gösta was suddenly gone. At first we thought he must be close by but soon realized that he wasn't. We began calling his name as we zigzagged our way through the crowd, trying to keep Julius close by us. It was like looking for a needle in a haystack trying to locate a small child in this sea of people and islands of exhibits that were several meters high and had no form of barricade.

I made my way over to the information desk and asked them to announce that a three-year old was missing but they told me they offered no such service.

"What?!" I asked, incredulous.

Instead of trying to reason with the pokerfaced information clerk I joined Bertil again and we wended our way through the herd calling out Gösta's name. It took around half an hour before Gösta reappeared, strolling casually towards us, oblivious to everything except the magnificent boats. He wanted to show us a boat he'd found that he thought we ought to buy.

A Weight

I have a boulder in my chest. It is huge and heavy and it prevents me from breathing deeply. The air I inhale doesn't reach down into the smaller airways in my lungs. I have had to learn to breathe shallowly.

The boulder also prevents me from stretching out my neck and lifting my face towards the heavens. I can no longer see the sky properly. I've learnt to accept this.

I can't listen to certain music without this boulder making its presence known and drowning out everything else. I've learnt to accept this.

Nor can I dance and prance around freely. I can hop a little but then this boulder drags me down with its weight. I've learnt to accept this.

Despite all this, there are times when I can forget that I have this boulder in my chest. Just for a little while each time. Life nowadays consists of gathering together these times when I can forget that I have a boulder in my chest.

***

Whenever Karin, Julius and I talk about our family, Karin always unhesitatingly mentions Gösta. This prompts two feelings in me.

On the one hand, it is as though Karin has conjured up a fantasy figure, something like her wanting to include her favourite soft toy in our family. But on the other, it is so obvious that Gösta belongs to our family and this means that it is impossible to see his death as a truth.

Sometimes I wonder which part of my life is the real part. Was it the life that I lived as mother of three or is it the life I've been living since 26 December 2004.

Maybe I'm just dreaming.
When will I be allowed to wake up?

***

The tsunami children are beginning to forget their siblings.
"It happened when I was little."
"Mamma, I hardly remember my brothers and sisters."
That's how it can sound.

***

Last night I breast fed Gösta. He was so big that he thought it was embarrassing to be breast fed so I put a thin blanket over him so that no one could see.
I held him tight.

Fifteen Years Ago

It is fifteen years since Bertil and I went to visit our friends in their apartment outside Stockholm. On the sofa in their sitting room there was a small bundle. Their daughter was only five days old.

Today is her fifteenth birthday. I should perhaps be thinking about how quickly time passes and that I haven't changed at all in these fifteen years even though this girl has grown from infant to young woman in that time. Such a huge difference for her and such a small one for me. But it isn't like that.

When I think back on that day when we stood admiring their newborn, when I was expecting Julius, it feels like an eternity ago. But it has nothing to do with chronological time – the past can feel so remote regardless of how many minutes or days have actually passed. Distance is not necessarily measured in terms of time, but also in terms of light years or centimeters.

Good-bye Mr. Muffin

My parents gave Gösta a book for Christmas in 2002. Inside the cover it says, "To Gösta, Christmas 2002, from Granny and Grandpa." The book is about a guinea pig who dies and it discusses what happens at death. It is written to help children understand death.

Why did Gösta get this book from my parents?

Does Anyone Want to Change Places with Me?

I've put the following announcement onto a website where you can advertise items you want to swap with others, "I'll take over your life if you'll take mine. I'll gladly take on your daily life, be it in a destructive relationship with your partner, drug addicted children, housing problems, unemployment, financial difficulties, whatever. There is just one condition: all your children must be alive. Then I'll gladly swap with you. I have a lovely house with a beautiful garden, good job and stable economy. I've separated amicably from my partner and have two wonderful children. I have no problems – but there is one thing. My son Gösta is dead. Do you want to swap?"

Karin is Sick

It is five o'clock in the afternoon. Karin is sick and bored. She has been home from school for five days now. She's been watching video films and playing computer games but now I've dug out an old video and have just put it into the player.

The film shows the last episode of a children's Christmas show and then it says that there are only two days left until Christmas. Swedish television wishes all the children a very happy Christmas – even Gösta, who was still alive at that time.

I've also taped Babar the Elephant. Karin is lying on the sofa waiting for the film to start up. She must have seen this film a hundred times already and I know the music from it like the back of my hand. When I hear it, I am overcome with old feelings and thoughts that used to mean nothing and now mean everything. I can see Julius and Gösta sitting on the sofa where Karin is now, waiting eagerly for the film to start.

Idol

Amanda has just sung Leonard Cohen's 'Hallelujah' on the TV program Idol and Kersti, who is the first in line to give it points is dumbstruck. Instead, she rises from her chair and starts slowly clapping her hands and then she is joined by the other two members of the jury.

Karin calls from the sofa that this is the music from Shrek. And then the pieces all fall into place. In my mind I see the cinema where we sat, Gösta beside me eating popcorn, while Cohen sang this song and the film played.

Amanda's song had pierced my heart but I didn't understand why and was able to hold back the feelings – it would be ridiculous to burst into tears because of a song sung on Idol!

On Sunday morning I am woken by my own longing for Gösta. My nose is swollen and eyes wet. I lie still until the dreams have ebbed away. Then I get up and go out to the kitchen, and while I wait for the water to boil, my eyes rest on the photo of me with all three children. Gösta is leaning against me with his arm around my waist.

When I look at the picture, I don't see how things were at that moment – I see how things are now. The image is perfect. We are standing on a hillock against a backdrop of dense jungle, a glassy lake, outcrops of rock. The three of us are framed in the foreground by a lacework of low-hanging branches with thick leaves.

I am on the left, Gösta is beside me with his head against my shoulder and Julius is on his other side, close enough that I can reach his neck with my hand. Julius is almost as tall as I am and Gösta is already up to my chin. Karin is at the edge of the picture, taking a step sideways. Her face it turned towards the camera and she is wearing the dress that saved her life. This is the picture of me with my three children – now and for ever.

As I look at it I feel the anguish welling up inside me and escaping from my mouth in a strange bleat. I don't fight it. I lean against the sink and let it gurgle up from within.

This process has a power that reminds me of giving birth. It is an unstoppable force of nature that comes from deep inside my body. I am unable to curb the primal cry that gushes up. My body has a life of its own and I have to obey, however unwillingly.

And then my mind starts playing tricks on me. I hear the Hallelujah song playing again and again in my head. I grab the kitchen towel and bury my face in it, then let it hang at my side, only to bury my face yet again. My only witnesses are the walls around me.

When the worst of it has eased, I pick up the phone and tell my friend that I won't be able to attend their party in the afternoon.

Holding Back

I would have liked to go home early but I stayed and listened to the band while my friends danced like madmen around me. They tugged at me, "Come on chicken, come and shake a tail feather!" I complied by swinging my hips a bit and moving my arms in time to the beat and grinning.

But at last I'm sitting on the night bus home. This damnable journey has taken more than two hours already and the evening is catching up with me. I'm angry with myself for not having left the party earlier and irritated by the fact that the trip is now taking so long but my anger suddenly dissolves into desperation; I feel like the clown who has made everyone laugh while she herself is so miserable.

I hold it in until I'm off the bus and then I let the tears come. It's almost three in the morning and no one can see me. But I see in front of me the forced smiles I put on a few hours earlier – the way I acted a role. As soon as anyone had looked at me I had pulled my lips into a grin as if to say, "Oh yes, I'm having such a good time!" And as soon as they looked away my mouth fell back into its normal position. Pathetic.

I feel like a howling child. But no one is allowed to see me like this. I'll only show this to ... no one.

Along the four hundred meters back to the house I bawl and feel the frustration of not permitting anyone to see this side of me.

Friends

Two friends are sitting in a cafe. They haven't seen each other for a while. They are sitting at a round table by a large window. Outside dusk is gathering and large snowflakes are floating softly down past the window. The cobblestones along the narrow street are almost completely covered now but the snow lifts in swirls around the feet of passers-by. The shop windows along the other side are twinkling cheerfully with Christmas lights.

A candle is burning on the table in the cafe and beside it are two cups of untouched cafe latte. The two friends have so much to talk about that they've forgotten all about their coffees.

"Felicia frightened the life out of me the other day," says Katarina. "She trotted off to the shop and didn't come back for ages. You know how it is – you try to tell yourself there's nothing to worry about but you can't stop worrying anyhow. In the end, I went down to the shop to find her and there she was. She'd just bumped into some friends and was standing talking to them. When I came tearing in she looked at me with one of those 'What's the problem?' looks. In fact, she hadn't been gone that long at all. It was just me being neurotic."

Katarina laughs at her own silliness. But Mia doesn't laugh.

"I know just what you mean. It can drive you crazy not knowing where your child is, even for a second. I remember when I couldn't get hold of Elias. He was with a friend and their family so I knew he'd be gone for a while. But I needed to ask him something, so I rang his friend's parents but they didn't answer. I tried their home number and their cell phones but there was no reply. It was horrible. I knew he was with them but it was terrible not being able to get hold of anyone when I wanted to."

The two friends continue discussing how parenthood inevitably means worry about one's child. These two women first met through their children but their friendship has deepened and they can share many feelings with each other now.

On the other side of the cobbled street is another cafe. There are two women sitting in there as well. Like Katarina and Mia, these two friends also have much in common through their children. They are also sitting by the window but their window is smaller and two candles are flickering on their table. And they two have two cups of untouched cafe latte standing in front of them. They are talking about their children.

"They found Amanda at the temple," says Åsa. "She had been carried there by some Thais. They called her the temple girl. When she arrived there she was apparently the only non-Asian and everyone in the temple had been fascinated by her long blond hair and pale skin."

Åsa picks up her cup in both hands to warm them. She shivers.

"It must have felt good that she was found so quickly. It felt like aeons until they found Gösta," says Ann as she picks up the spoon on the table and twiddles it in the froth on her coffee.

"There is such a difference between understanding something and knowing it for sure."

"Yes, that's true. We knew quite quickly. And we got to fly on the same plane home. I could bring Amanda home."

"We met Gösta at the military airfield four months later."

"Was Gösta cremated?"

"Yes."

"I couldn't bring myself to have Amanda cremated. Just the thought of incinerating her body felt horrific."

"I can understand that. But we had Gösta cremated. His remains fit into a small urn. The body that was left wasn't him anymore anyhow, and the thought of what happens to a body when it's buried..."

"True. And it is only a shell."

Åsa sips her coffee and looks out of the window. It has stopped snowing and the plaster-works on the buildings on the other side are glowing in the light of the street lamps. She catches sight of two women of about her own age sitting in a cafe across the street. They are leaning forward and talking intently. She sees how one of the women then slaps her thigh, leans back in her chair and laughs. They seem to be enjoying themselves those two, on the other side, thinks Åsa to herself.

Then Katarina looks up and notices Ann and Åsa, just across the road, sitting as she is with her friend, chatting.

"Look over there," she says to Mia. "I'll bet they're sitting there talking about their kids too. How much attention should kids get really?"

"Yeah, I'll bet you're right," says Mia. "Doesn't it look cosy with all the snow outside and the candle on the table in there. I hope they're having a good time too."

Another Christmas Approaches

A famous South African jazz musician is playing the piano, and although it isn't my favourite kind of music I find myself jiggling my foot in time to it.

I begin imagining the Christmas to come; we are all at Auntie My's house with our remaining relatives. We've all helped prepare the food and decorate My's home. It's so Christmassy.

I'm glad for Julius' and Karin's sake that they'll be celebrating Christmas with their cousins, whom they love.

Christmas arrives and everything is calm. I can see how I shall be gathering all our things in the evening in readiness to leave for home and how the children protest and want to stay longer. So I let them stay and I go home alone for an early night.

The thought of 26 December is commanding. I'll be sitting there alone, listening to music and my defences will crumble and I'll capsize into a heap.

I focus on the jazz musician's back instead and try to resist but I can feel my eyes brimming. Must I start sobbing again?

Christmas is approaching and I have to enter the tunnel.

Karin Remembers

"At school today I lay down on the sofa and cried. I was so sad."

"Oh Karin, why's that?"

Karin is standing by the table laying out the cutlery for supper while I prepare the food.

"I was thinking about Gösta."

"Were you? Were you all on your own?"

"No, there were other children there from the parallel class."

"Did you tell them why you were so sad?"

"Yes."

"I told them how Julius had stood holding us and gripping onto a tree and how we couldn't find Gösta and how we ran up to the jungle because a new wave was coming and how we ate squid and had to stay there all night."

"Did you tell them that we sat on banana leaves too?"

"No, but I told them that you got dragged under a car and had lots of sores on your back after."

Karin goes quiet for a while and looks thoughtful, as though she's trying to work out what she's feeling.

"Then I told them about Gösta and I told them what he looked like and that he was a really kind big brother. I just get sadder and sadder when I think about it so I told them that I was going to go and read a book, and then I got up and went and got a book and looked at all the pictures."

Christmas Rush

There is so much to do on the day before Christmas. Early in the morning I begin wrapping the last of the presents so that at least that will be completed. Then I have to get Kasper's food bowl because I'm going to collect him before I head off to My's later on. I put the kettle on, take out some paper and a pen and sit down at the table.

What do I need to buy? Mustard and crème fraîche for the sauce, sour cream for the herring. Is the liquor store open? Oh well, we have enough anyhow.

This is going to be a truly modern Christmas at My's house, with all the ex-partners and new partners, old relatives and new. It'll be alright. Tomorrow morning I shall eat breakfast with Bertil, where Karin and Julius are and after that we shall go to Auntie My's and Christmas. When I've completed my shopping list, I pack what I need and get into the car. It isn't even ten o'clock yet so I have plenty of time to do what I need to before going to My's to make Christmas food together with her and our father.

I am almost alone at the florists. I just need to buy a flower to put on Gösta's grave before I pick up Kasper. We have found a good home for him in Enköping and he lives there now with his new owner, Karla, and four dog friends. It was impossible to have him home when I began working full-time so we have agreed that we can borrow him whenever we want to. We usually bring him home during the holidays when Julius and Karin are with me. Karin has been begging that Kasper celebrates Christmas with us this year.

One of the shop assistants asks if I need some help and I immediately decide on a rose. I explain to the young woman who is serving me that it will lie outside, on my son's grave. It is as though I am telling this to myself, for the first time: "Thank you, I would like to buy a red rose to put on my son's grave."

The young woman picks out a beautiful rose and asks if I would like her to wrap it in paper. I nod and stop breathing, but the tears will not be held back. I pay, take the rose and hurry out.

It is tranquil and silent by the grave. I kneel down and start to tidy up around it. I brush away the dead leaves and pick up a plastic vase full of wilted flowers. I wonder who has left them here? I remove the burnt out jars with candles inside and arrange the shells and stones. Then I unwrap the tissue paper from the rose and place it carefully onto the grave. I rise, still looking at the grave. The rose is from me to Gösta.

The air is mild and a mist is hanging over the graveyard and although everything is toned in grey, it is peaceful and beautiful. Below the cemetery I can see the road snaking away over the dark brown fields. Ahead of me, it is as though I can see two small boys on their way home from school, as if the school bus has dropped them by the church so they have to walk the final length; they are taking brave strides and they have one another.

I wrench my eyes from the road and look down at the grave again. No, it is here that Gösta lies and I have laid a red rose on him.

I gather together all the scrap from around the grave and start off for the car again and the Christmas rush. Mustard, crème fraîche, sour cream …

***

This is so painful.

Imagine that your child has died, I mean for real. Imagine that! If you have more than one, pick one of them. What is that child's name? Think of that child and imagine that he or she is no more.

I know; it's unimaginable. It is too agonizing to imagine oneself into my situation for real. If you can, then you'll probably only manage for a short while and then it'll be too painful. Maybe this is a means of self-preservation.
I don't want to torture you anymore. You don't need to think about it anymore. You can stop now.

But the feelings you experienced, if you managed, only for that short moment – those are the feelings that I live with constantly.

Epilogue
November 2009

And what is life like now?

More time has passed, life continues, new experiences paint over old ones. I meet more and more people who never knew Gösta and I slowly work my way forwards.

But what happens to Gösta? Does he glide backwards then? Where is he in all this? What does it feel like now to think about him? Easier? No, not at all.

I may find it easier to distract myself from thinking of him with all that's happening in my life now. But the drop from joy to sorrow seems to be higher than ever. There is just as much that I hate and I feel defenceless.

It isn't schizoid. It's not that I have two personalities but more that I live on two levels, in two separate rooms.

One level is up, now, real. But below it is another world and it is abhorrent. If I was stuck down there all the time, I wouldn't last. Time stands still down there; Gösta is there, but only as a mirage, an agonizing feeling of longing and sorrow. It is as painful as ever, as painful as it has been since the moment I realized that he was gone and that a part of me had died with him. That will never change.

But the distance between these two levels has increased. At the upper level, Gösta is gradually being displaced. Other children have caught up with and overtaken him.

In only another year or so Julius will be fully grown. He has become so tall and his voice is no longer that of a child. His round cheeks have become angular and he has huge feet. He is becoming more and more independent and is forging plans for his future. He is in Australia now and will be there for almost a year. Behind him he leaves a little brother who will never be more than nine years old – a brother who used to follow his big brother's every move but who has now got left behind.

Karin is no longer a little sister. She is now in the same class as Gösta was and she is nine years old. The way she plays, her gestures and movements remind me of Gösta. They would have had such fun together. Now she is the same age as him.

New memories. New experiences.

Every now and then I get knocked off balance and become consumed with the fact that Gösta is dead and is nothing more than a memory.

There is usually a trigger – something that reminds me of a past life. It could be an aeroplane that's just lifting for Singapore. It could be a poem, a song, a play, a reminder of a life that once was.

I'm often asked what it's like now that Julius has gone to Australia, whether I think it's awful to have him so far away for such a long time. Of course it is. But I'm used to not having all my children with me. It is several years since I last saw Gösta.

I recently chatted to Julius in Australia and afterwards I was so irrationally angry. In my childishness I still try to pretend that Gösta is somewhere else, on the other side of the world, and that we'll soon see each other again. When I speak to Julius it brings it back to me that this isn't true.

Gösta isn't somewhere else. He is dead and I shall never see him again. It's impossible to reach him, ever. He doesn't exist.

So it isn't so difficult to have Julius on the other side of the world. He is still here, round the corner, he can be reached. Gösta can't.

Those words have lost none of their poignancy.

It is still just as frightening and foreign. It is monstrous.

Gösta is dead.

www.ingramcontent.com/pod-product-compliance
Lightning Source LLC
Chambersburg PA
CBHW021139090426
42740CB00008B/852